And while there were many reasons why Peter's work proceeded slowly, among them were his knack for keeping things entertaining. Shortly after Watson and Crick's discovery of DNA for example, he fabricated a letter of invitation from his father to Crick, requesting Crick's presence at an upcoming conference on proteins at Caltech. "Professor Corey and I want you to speak as much as possible during the meeting," the impostor Pauling said to Crick in the fake letter, even urging him to consider lecturing at Caltech as a visiting professor. Linus Pauling had appeared to sign the letter himself, his signature skillfully forged. The letter proved so convincing that Crick actually replied, accepting the invitation to speak at the conference.

But before long, it became apparent that the entire communication was, in fact, a practical joke, mostly because Lawrence Bragg, the director of the laboratory where Crick himself worked, was scheduled to speak at the conference in the same time slot that the fake letter had proposed for Crick. Were it not for this glitch the deception might have gone even farther, since upon seeing his son's forgery, Linus himself was almost convinced that he had written the letter and simply forgotten about it. Ever a stickler for the details however, Pauling noticed a grammatical error in the document that he would never have made, and deduced the letter as having been authored by his mischievous son. For this transgression, Linus subtracted a five-pound fine from the $125.00 check that he sent to Peter each month.

VISIONS OF
LINUS PAULING

Other Titles by the Editor

The Pauling Catalogue

VISIONS OF
LINUS PAULING

Editor

Chris Petersen
Oregon State University, USA

NEW JERSEY · LONDON · SINGAPORE · BEIJING · SHANGHAI · HONG KONG · TAIPEI · CHENNAI · TOKYO

Published by

World Scientific Publishing Co. Pte. Ltd.
5 Toh Tuck Link, Singapore 596224
USA office: 27 Warren Street, Suite 401-402, Hackensack, NJ 07601
UK office: 57 Shelton Street, Covent Garden, London WC2H 9HE

Library of Congress Cataloging-in-Publication Data
Names: Petersen, Chris (Christoffer), editor.
Title: Visions of Linus Pauling / editor, Chris Petersen, Oregon State University, USA.
Description: New Jersey : World Scientific, [2023] | Includes index.
Identifiers: LCCN 2022019486 | ISBN 9789811260759 (hardcover) |
    ISBN 9789811260766 (ebook) | ISBN 9789811260773 (ebook)
Subjects: LCSH: Pauling, Linus, 1901-1994. | Chemists--United States--Biography. |
    Biochemists--United States--Biography.
Classification: LCC QD22.P35 V57 2023 | DDC 540.92 [B]--dc23/eng20220822
LC record available at https://lccn.loc.gov/2022019486

British Library Cataloguing-in-Publication Data
A catalogue record for this book is available from the British Library.

For any available supplementary material, please visit
https://www.worldscientific.com/worldscibooks/10.1142/12977#t=suppl

Desk Editor: Shaun Tan Yi Jie

Typeset by Stallion Press
Email: enquiries@stallionpress.com

Printed in Singapore

About the Editor

Chris Petersen is Senior Faculty Research Assistant at the Special Collections and Archives Research Center (SCARC), Oregon State University Libraries and Press, USA, where he has worked since 1996. He is also the founder, editor, and publisher of *The Pauling Blog*, which has released original research on Linus Pauling nearly every week since its creation in 2008. In addition, he administers the SCARC oral history program and has led more than 200 interview sessions comprising in excess of 300 hours of content. He previously co-edited *The Pauling Catalogue*, a six-volume set describing the Ava Helen and Linus Pauling Papers, which are housed at Oregon State University Libraries.

Preface

By Chris Petersen

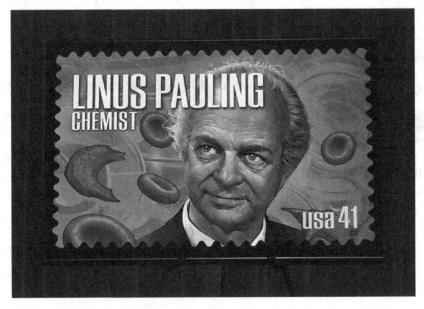

The Pauling "American Scientists" series postage stamp, as unveiled on the campus of Oregon State University, March 6, 2008.

All of the chapters presented in this book were originally published as blog posts for a project that, as I'm fond of recounting, was created to celebrate the release of a postage stamp. In March 2008 the United States Postal Service issued a series of stamps honoring American scientists, one of whom was Linus Pauling. A day before the series was released nationwide, a special early-issue event was held in New York

City and also in Corvallis, Oregon, which is home to Pauling's undergraduate alma mater and my workplace, Oregon State University (OSU).

By then my department, the OSU Libraries Special Collections,
had been home to the Ava Helen and Linus Pauling Papers for more
than 20 years, a time period during which we had been closely involved
with a great many activities related to Pauling and his legacy. In the years
prior to the Postal Service event, the only real promotional tools available
to us were word of mouth, university news releases, a very small spot
on our website, and an RSS feed that was likely subscribed to by only a
handful of people. Looking to try out something that might make a bigger impact, we created a WordPress blog — titled The Pauling Blog —
and devoted the first handful of posts to the stamp and the celebration.

Before too long, the shape of the project began to change quite
significantly. Excited by this new (to us) platform, we decided to move
beyond promotional activities and instead devote time and resources to
conducting original research that would explore particular aspects of
Pauling's life on a deeper level than one might find in a typical biography. From Spring 2008 on, we continued in this vein, consistently and
diligently, up into the pandemic, and can now boast of a resource that
contains more than 850,000 words and has been viewed more than 1.25
million times.

As noted, the Pauling Blog came to pass within the context of a
much larger Pauling-related program that was housed in Special Collections. Linus Pauling donated his papers to OSU in 1986, and did so
despite numerous offers from much larger and more prestigious repos-
itories. Ever the maverick, he instead chose the option that, at the time,
had neither space nor staff enough to process or provide access to what
was understood by all to be a very large collection.

There are multiple theories about why Pauling ultimately made
this decision, but for me the deciding factor was OSU's commitment to
devote substantial new resources to preserving his life's work. To this
end, a brand new department — Special Collections — was created
and outfitted with three full-time employees who were supported by

a large team of student workers. Naturally, much of the effort in those early days was focused on organizing the collection, but as time moved forward, attention increasingly turned towards the development of digital resources that would tell the Pauling story in unique and accessible ways. Core to this were several documentary history websites that teased out specific aspects of Pauling's life, contextualized them with a unifying narrative, and illustrated them with materials selected and digitized out of the Pauling Papers. The department also played a central role in organizing three major conferences devoted to Pauling, and likewise hosted a great many presentations delivered by a rotation of resident scholars as well as recipients of the Pauling Legacy Award, which was granted by the OSU Libraries every two years.

In 2011, Special Collections and the University Archives at OSU Libraries merged into a new super-unit called the Special Collections and Archives Research Center. One of the primary motivations for this merger was to leverage archival skillsets across a very large swath of materials, and partly as a result of this, the emphasis placed on Pauling activities was decreased. In this new circumstance the Pauling Blog took on added importance, and today it remains both the unit's primary form of Pauling outreach as well as the strongest lingering connection to the Special Collections era.

When I was approached by the editors at World Scientific with the idea of transforming components of the Pauling Blog into a book, I quickly realized that I would need to devise a strategy for distilling down a vast web-based project into a single print publication. As I puzzled over this more, I realized that it would not be possible to craft anything like a biography from our source material, as the content on the blog is both too detailed and too narrowly focused to jam into a book of about 300 pages. Instead, I chose to select pieces that, in my view, represented novel and significant contributions to our evolving understanding of Pauling's activities and era.

As such, it is important to emphasize that this is not a biography. Indeed, the book's first chapter finds us encountering Pauling when he is already in his mid-30s and working on what would become a master-piece of 20th-century science, *The Nature of the Chemical Bond*. But while others have investigated the makeup and impact of this book, to my knowledge none have discussed Pauling's actual experience of writing it. This is the vantage point that we take in our exploration of the text, a perspective that makes effective use of our proximity to Pauling's papers and also provides for a useful contribution to the scholarly understanding of the Pauling story.

Many other examples of this approach can be found throughout the book, including a detailed exploration of his relationship with his publisher; insights into two moments where he nearly died; additional insight into his actual passing; a close examination of his receipt of the Nobel Peace Prize; and deep dives into his clashes with, variously, the Senate Internal Security Subcommittee, the Soviet Academy of Sciences, and *National Review* magazine. Likewise included are the colorful history of the institute that bears his name, and the complicated biography of his second-born child. Lesser-known aspects of his scientific resume are also probed, including his work on anesthesia and quasicrystals, as well as the many patents that he pursued. Shorter pieces have been selected to shed light on his personality and to learn more about, for instance, the Pauling family's rather unorthodox Christmas traditions. Finally, to situate all of this within the context of a larger life narrative, I have included a biographical sketch that will hopefully fill in some of the details for those less well-versed in the Pauling story.

For more than 14 years, the Pauling Blog has worked as a collaboration between myself and a large team of writers, almost all of them students at OSU. Generally I have been responsible for generating topics and guiding students to relevant series within the collection, and then once their manuscripts have been completed, I edit them, select illustrations and publish them on the blog. The authors of pieces selected for presentation in this book have been noted with each chapter, but they represent only a fraction of the people who have contributed to the project's successes. The full roster of talented folks who have written for the Pauling Blog are as follows:

Jamee Asher, Marcus Calkins, Jindan Chen, Will Clark, Maddie Connolly, Carly Dougher, Anna Dvorak, Joseph Esposito, Melissae Fellet, Desiree Gorham, Melinda Gormley, Andy Hahn, Ethan Heusser, Madeline Hoag, Ben Jaeger, Rachel Koroloff, Adam LaMascus, John Leavitt, Miriam Lipton, Sarah Litwin, Luis Marquez Loza, Matt McConnell, Michael Mehringer, Jessica Newgard, Ingrid Ockert, Jaren Provo, Audrey Riessman, Shannon Riley, Olga Rodriguez-Walmisley, McKenzie Ross, Libby Runde, Trevor Sandgathe, Tina Schnell, Geoff Somnitz, Megan Sykes, Dani Tellvik and Ryan Wick.

As we developed our ideas for posts, time and again we leaned on canonical texts written by two of Pauling's biographers — Thomas Hager and Robert Paradowski. Both are giants of Pauling studies to whom we owe sincere thanks. We are similarly indebted to three department heads who have supported this project into its teenage years — Cliff Mead, Larry Landis and Natalia Fernández. Likewise, this book would not have come into existence without the encouragement of Shaun Tan Yi Jie, my editor at World Scientific.

One final note; a personal one. In many respects, mine has been a charmed life filled with love, friendship and inspiration. A great many people have supported me over the years, but none more so than Karen and Nora. This book is dedicated to them.

Biographical Sketch of Linus Pauling

By Chris Petersen

Pauling during his years at Oregon Agricultural College, 1918.

Linus Carl Pauling was born to parents Herman and Belle in Portland, Oregon, on February 28, 1901. He was the first of three children, preceding sisters Pauline (1902–2003) and Lucile (1904–1992). Herman Pauling was a pharmacist by trade, and in 1905 the family moved to Belle's hometown of Condon, in hopes that Herman's skills might find a market. Condon was a small farming community located more than 200 miles away from Portland and it was here that Linus began his schooling, up through the fourth grade.

In 1910 the family returned to Portland and, soon after, tragedy struck when Herman died of a perforated ulcer. Young Linus was immediately called upon to contribute to the family income, and in the years that followed he worked a number of odd jobs, including delivering milk from a horse-drawn wagon and setting up pins at a bowling alley. Later on in his adolescence, Pauling found employment as a drill-press operator, work that came easily to him and paid well. He was encouraged by his mother to consider a career at the machine shop, but his ambitions lay elsewhere. At the age of 12, Linus had been shown a few basic chemical experiments by his close friend Lloyd Jeffress. A bright and introspective boy, Pauling was immediately taken by this introduction to the world of chemistry and soon understood that his deepest passion was science.

In 1917 Pauling enrolled at Oregon Agricultural College (OAC) in Corvallis, despite the fact that he remained two classes shy of graduating from Washington High School in Portland. At the time, OAC did not require a high school diploma for admission, meaning that, at the tender age of 16, Pauling was allowed to begin his collegiate studies. Initially unsure of his fitness for college life, Pauling quickly recognized that his skillset and acumen had placed him far ahead of his peers. Majoring in Chemical Engineering — OAC did not offer a Chemistry major during this period — and working full-time to make ends meet, Pauling excelled as a student and flourished socially, in part through the contacts that he made as a member of the Delta Upsilon fraternity.

To help fund his academic pursuits, Pauling spent his summers working in the field as a pavement inspector for the Oregon Department of Transportation. Near the end of the summer in 1919, he learned that his mother had been forced to spend his savings to support herself and her two younger daughters, and as a result he would not be able to return to Corvallis for his junior year. He remained with the paving crew into the fall, until receiving an offer from OAC that he come back as a full-time instructor. Still just 18 years old, Pauling readily accepted this appointment, teaching for the remainder of the academic year. The "boy professor" re-enrolled as a student the following fall, while continuing to teach part-time. It was in this capacity that he met Ava Helen Miller, a Home Economics student who had signed up for one of his introductory chemistry courses. The two quickly fell in love and by the spring of 1922 Pauling had proposed marriage. Both of their mothers disapproved of this idea, and after completing his OAC degree, Pauling left for graduate studies at the California Institute of Technology (Caltech) while Ava Helen remained in Oregon. Early the next summer the couple were wed, and Ava Helen moved to Pasadena with her husband.

Caltech proved to be an ideal location for Pauling. Under the guidance of mentor Roscoe Dickinson, Pauling mastered the techniques of X-ray crystallography, which he continued to advance in subsequent years to solve the crystal structures of a great many compounds. He completed his PhD in 1925 and was immediately offered funding by the Institute to stay in Pasadena. Crucially, Pauling also received a Guggenheim Foundation grant to study in Europe during the 1926–27 academic year. Leaving one-year-old Linus Jr. behind in the care of Ava Helen's mother, the couple spent the year abroad, with the elder Linus focusing intently on absorbing as much as he could about the emerging field of quantum mechanics. In 1927 Pauling returned to a faculty position at Caltech and a period of momentous work, during which he began to apply the new physics to the classical understanding of how atoms form into molecules. In 1931 the first in a series of papers bearing the prefix "The Nature of the Chemical Bond" appeared in the *Journal of the American Chemical Society*. Over the next two years, six more of these papers were published, their findings summarized and expanded upon in a 1939 book, also titled *The Nature of the Chemical Bond*. The work was quickly recognized as being of major import, and certain ideas that it put forward — including the theory of hybrid bond orbitals and a novel electronegativity scale — proved fundamental to the development of modern chemistry. In 1954 Pauling would receive the Nobel Chemistry Prize for this body of "research into the nature of the chemical bond and its application to the elucidation of the structure of complex substances."

In the 1930s, Pauling increasingly shifted his interests toward biological topics, in part inspired by the Rockefeller Foundation's desire to fund projects focusing on the "science of life." Pauling's initial concentration was the hemoglobin molecule, which he studied using the nascent technique of electron diffraction. In time, Pauling expanded this program to include groundbreaking investigations in immunology and the structures of other complex proteins. Two major breakthroughs came about in the 1940s, first with Pauling's discovery of the alpha helix — a major structural configuration found in many protein molecules. (A rushed triple-helical proposal for DNA published in February 1953 proved far less successful.) Pauling also led a group that experimentally proved that sickle cell anemia is a molecular disease, the first condition to be so described. Likewise, as an outgrowth of his immunological investigations, Pauling showed that antibodies and antigens bond as a result of precise complementarity of shape, a trait called biological specificity. It was during this period as well that Pauling published the first edition of his seminal textbook, *General Chemistry*, thus beginning a long and fruitful relationship with the W.H. Freeman & Co. publishing house.

As with many other scientists, the war years forced Pauling to pivot from his established research agenda. Scientifically, Pauling turned his attention away from protein structures in favor of government contracts to improve rocket propellants and secret inks. Pauling was also charged with the development of an oxygen meter for use in aircraft and submarines, and artificial blood that might aid wounded soldiers in the field. The conclusion of the war proved to be of greater consequence to Pauling's story. Shaken by both the devastation and implications of the atomic bombings of Hiroshima and Nagasaki, Pauling quickly felt drawn by a moral imperative to speak out against the proliferation of these new weapons of mass destruction.

Though generally interested in many issues emergent during the nuclear age, Pauling eventually found a niche as a leading voice against the testing of nuclear weapons in the atmosphere. Pauling was among the first to forcefully argue against the practice of above-ground test detonations, pressing his belief that the radioactive fallout unleashed by these tests — ever larger and increasing in number as the Cold War advanced — was contaminating the Earth in ways that would seriously damage human health. In 1957 he and Ava Helen decided to focus their anti-nuclear activism in the form of a petition that would be circulated around the world with the aim of affirming scientific consensus against above-ground nuclear testing. Ultimately signed by more than 11,000 scientists across the globe, the "Appeal by Scientists to Governments and People of the World" was formally presented to the United Nations in January 1958.

Pauling's bomb test petition did much to amplify the debate around weapons testing and, in the estimation of many, helped prompt the Partial Test Ban Treaty agreed to by the United States, the Soviet Union, and Great Britain in June 1963. Pauling received his second unshared Nobel Prize that December, "for his fight against the nuclear arms race between East and West." And though this and other decorations cemented his reputation as a hero to many, his outspokenness during a tense period in history also exacted a significant personal cost.

As his profile rose, Pauling was routinely dismissed as a Communist sympathizer by critics in the media and government. In 1952 his passport was temporarily revoked, and in 1960 he was twice called before the Senate Internal Security Subcommittee with orders to reveal the names of those individuals who had helped to circulate his petition. Faced with threats of imprisonment for contempt of Congress, Pauling refused to comply with the demand, and ultimately the subcommittee stood down. While he won that particular battle, episodes of this sort led to mounting criticism from conservative media commentators, some of which resulted in a spate of libel lawsuits filed by Pauling in the early- to mid-1960s. Pauling also gradually lost his footing at Caltech, as trustees and other donors came to take a dim view of his non-scientific pursuits. Pauling had been appointed chair of the Institute's Division of Chemistry and Chemical Engineering in 1937 and oversaw a period of significant institutional growth until 1958, when growing calls that he step down finally reached a crescendo. Following that, Pauling's office and laboratory space allotments were reduced. The final straw came with the Institute's decision not to formally acknowledge Pauling's receipt of the Nobel Peace Prize. By the time he had returned from the award ceremonies in Oslo, Pauling had resigned from Caltech, his institutional home of 40 years.

What followed was a period of wandering that began with a move to a think tank, The Center for the Study of Democratic Institutions, located in Santa Barbara, California. Though invigorating in certain respects, the center lacked scientific apparatus of any kind, and before long Pauling had moved again, this time to the University of California, San Diego. After two years in La Jolla, Pauling relocated once more, to Stanford University. It was during these nomadic years that Pauling developed what would become a famous fascination with vitamin C. Spurred by a chance encounter with a biochemist named Irwin Stone, Pauling became increasingly convinced that chronically low concentrations of ascorbic acid in the bloodstream were a major contributor to suboptimal health for humans, who are among a small handful of the Earth's species incapable of synthesizing their own vitamin C. As the 1960s moved forward, Pauling applied this lens to concurrent interests in mental health, coining the term 'orthomolecular psychiatry' to describe the treatment of disorders like schizophrenia with appropriately dosed vitamins and minerals. Subsequent investigations were circulated among the public in books, including the best-selling *Vitamin C and the Common Cold* and a later volume titled *Cancer and Vitamin C*.

Pauling's vitamin C work was looked upon with a skeptical eye by much of the medical community as well as many scientific colleagues. As the controversy surrounding his dogged advocacy of orthomolecular medicine mounted, and as his own interest continued to grow, Pauling increasingly felt a need for new institutional support. In 1973 he began a new chapter by co-founding what would become the Linus Pauling Institute of Science and Medicine (LPISM), originally located in Menlo Park, California, and later in neighboring Palo Alto. For the remainder of Pauling's life, his namesake institute pursued an often sprawling research agenda, a core piece of which remained connected to the original focus on vitamin C. LPISM did so amidst high-profile setbacks to its research agenda, chronic issues securing funds, and a series of personnel conflicts that resulted in costly periods of litigation. Pauling's burden was made all the heavier in December 1981, when Ava Helen succumbed to stomach cancer at the age of 77.

Linus would outlive his beloved wife by another 12 and a half years. During that time, he continued to speak out forcefully for peace and against nuclear weapons during the Reagan years, and again against the Persian Gulf War in the early 1990s. Scientifically, his vitamin C interests came to include investigations on its possible benefit for heart health. Pauling also engaged deeply with theoretical studies during this time period, focusing in particular on quasicrystals, transition metals, and his own theory of nuclear structure, and spending many days in solitude, working through calculations at his secluded ocean-side home near Big Sur. At the end of 1991, Pauling was diagnosed with rectal cancer, an illness that gradually advanced over the next two and a half years. He passed away at his home on August 19, 1994, aged 93 and survived by his children Linus Jr. (b. 1925), Peter (1931–2003), Linda (b. 1932), and Crellin (1937–1997). In a career spanning more than 70 years, Pauling published over 1,100 articles in the scientific and popular press, and was granted 47 honorary doctorates. Among many other awards, Pauling remains the only person to have received two unshared Nobel Prizes.

Contents

1

Writing *The Nature of the Chemical Bond*

By Andy Hahn

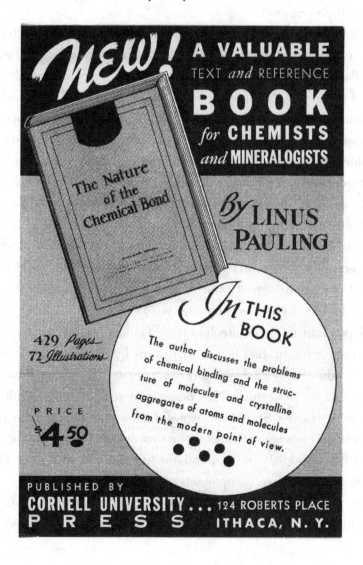

Linus Pauling's *The Nature of the Chemical Bond*, first published in 1939, was the product of over two decades of diligence, sacrifice, and collaboration that involved a broad range of actors including Pauling's family, research assistants, professional colleagues and a variety of institutions. Pauling's prefatory remarks to the book — "For a long time I have been planning to write a book on the structure of molecules and crystals and the nature of the chemical bond" — give an indication of the extent to which it had been a long-term objective for Pauling, despite his being only 38 years old.

Pauling's application for a grant from the Carnegie Institute in February 1932 provides a more detailed affirmation of his ambitions. In it, Pauling relayed how his undergraduate studies of crystal structures at Oregon Agricultural College between 1917 and 1922 had laid the foundation for his current work by bringing him into contact with contemporary questions in structural chemistry. As a graduate student at Caltech, Pauling had then begun to search for answers to those questions by delving into the newly developing field of quantum mechanics.

In 1926, Pauling and his wife Ava Helen — with the support of a Guggenheim Fellowship — left their one-year-old son, Linus Jr., with Ava Helen's mother in Portland and traveled to Europe to study quantum mechanics at its source. There, Pauling deepened his understanding and immersed himself even more by beginning to apply the new physics directly to chemical bonding.

Upon returning to Caltech in 1927, Pauling began to seek funding to continue what he had begun. Let down by the National Research Fund, Pauling supported his work with grants from Caltech and the National Research Council. This money allowed him to hire a full-time assistant, J. Holmes Sturdivant, who focused on X-ray crystallography and continued to work with Pauling for many years. Pauling also brought aboard Boris Podolsky for nine months to assist him with some of the more detailed technical components of his investigations.

In 1932 Pauling expressed a hope that, with help from the Carnegie Institute, he could expand his project by bringing in more assistants and purchasing equipment including an "electric calculating machine," a "specialized ionization spectrometer," and a microphotometer. The Carnegie Institute was not interested. Luckily for Pauling, the Rockefeller Foundation did come through with a general grant of $20,000 per year over two years, to be split between the physics and chemistry departments at Caltech. This allowed Pauling to keep Sturdivant on staff while adding three others — George Wheland, Jack Sherman, and E. Bright Wilson Jr. — to his research team.

The scramble to secure funding and recruit new people into the lab came amidst the publication of Pauling's first four "Nature of the Chemical Bond" articles in the *Journal of the American Chemical Society*, proof positive that Pauling's efforts were bearing fruit. Once the funding was secured and Sherman and Wheland had begun producing results, Pauling wrote — with Sherman and Wheland as co-authors — three more "Nature of the Chemical Bond" articles the following year, appearing this time in the newly established *Journal of Chemical Physics*. Wheland also worked with Pauling on a monograph discussing the application of quantum mechanics to organic molecules. The assistant finished his part of the book by 1937, but Pauling never got around to his portion. Instead, his desire to write a book-length treatment of chemical bonds began to move closer and closer toward center stage.

In order to keep the funding coming in through the lean years of the Great Depression, Pauling was compelled to follow the lead of his patrons, the Rockefeller Foundation. Warren Weaver, Director of Natural Sciences for the Foundation, told Pauling in December 1933 that the organization was "operating under severe restrictions" and that funding would go to projects "concentrated upon certain fields of fundamental quantitative biology." Evidence that Pauling's work had "developed to the point where it promises applications to the study of chlorophyll, hemoglobin and other substances of basic biological importance" was key to his potential receipt of continued dollars.

A pledge from Caltech's chemistry department to continue pursuing the line of research suggested by Weaver helped Pauling to secure funding for the following year. A three-year commitment came after that, providing the Caltech group with a reliable source of support into 1938. Pauling thanked Weaver in February of that year for his direction, writing, "I am of course aware of the fact that our plans for organic chemistry not only have been developed with the aid of your continued advice but also are based on your initial suggestion and encouragement; and I can foresee that I shall be indebted to you also for the opportunity of carrying out on my own scientific work in the future to as great an extent as I have been during the past six years."

Secure funding allowed Pauling to maintain a research group consisting of graduate students and post-doctoral fellows. In his preface to *The Nature of the Chemical Bond* book, Pauling would express gratitude to several of these individuals, including Sherman and Sturdivant. Another, Sidney Weinbaum, would earn his doctorate under Pauling and continue on afterwards, helping Pauling with quantum mechanical calculations and molecular structures.

Among others referenced in the preface, Fred Stitt worked as a research fellow with Pauling and assisted him in teaching his graduate course on the applications of quantum mechanics to chemistry — an exercise, no doubt, that helped shape Pauling's own thoughts on the subject. Charles Coryell also worked as a research fellow on the topic of magnetic susceptibilities, which were central to investigating chemical bonds. (Coryell later helped Pauling to construct a magnet for the Caltech labs, based on one already in place at Cornell.) A different collaborator, Edwin Buchman, was self-supporting due to royalties that he was earning from his synthesis of vitamin B1. Buchman was eager to help, telling Pauling in May 1937 that he would assist "on any problem in which an organic chemist could be useful and for which extra space could be had."

*

Though focused and motivated to compose a book-length treatment of the chemical bond, for a time Pauling seemed destined not to complete it. Burdened, in a sense, by his own and others' rapid advancements in understanding, early attempts at what would become *The Nature of the Chemical Bond* quickly went out-of-date if they were even briefly set aside.

A window opened at the end of 1936, when Pauling received offers to serve as visiting fellow at two different institutions on the East Coast. One came from the Institute for Advanced Study in Princeton, New Jersey, and the other from Cornell University in Ithaca, New York. Cornell sought to enlist Pauling as George Fischer Baker Lecturer for the year, a prospect that seemed to include both the time and structure necessary to write his book, as every year's lectures were followed by a publication.

From there, Pauling worked to discern exactly how his writing of *The Nature of the Chemical Bond* could fit in with the Baker Lectureship. In November 1936, he asked Jacob Papish, who was arranging the fellowship, if an expanded text based on his lectures was possible and how much the book might cost. Pauling wanted the price to be set as low as possible to promote a "good sale," and based his expectations on the one cent per page cost of previous books published in the Baker series. Royalties were also of interest as Pauling was already planning additional editions and expansions of his yet unwritten book. Papish welcomed Pauling's idea and suggested (very correctly, as it turned out) that his book would be one of the most successful of the series. However, all royalties for the first edition would go to the Cornell University Press, with income for any subsequent editions belonging to Pauling.

These terms were agreeable and, with everything seemingly arranged by December, Pauling only needed approval to take leave from Caltech. The death that previous June of the Institute's chemistry head A.A. Noyes created some hesitation in the minds of those around Pauling; as he told Papish, "the authorities of the Institute" questioned whether it was appropriate for him to take leave in a season of change. But the matter was resolved quickly and Pauling began to plan for his trip.

Initially Pauling hoped that his whole family, including Ava Helen, Linus Jr. and two other children, Linda and Peter, could join him in

Ithaca, where they would all stay together in a house. But the family was growing and, in June 1937, Ava Helen gave birth to the youngest Pauling child, Crellin. His attempts to convince Ava Helen otherwise having failed, Pauling told Papish a month before the birth that it would probably only be him showing up. Ultimately Ava Helen did wind up joining her husband on the train and staying with him for about a month, leaving the children and their dog Tyl in the care of Lola Cook, an employee of the Paulings who lived with the family and assisted with childcare and household chores.

Preparations related to Pauling's Caltech responsibilities were also necessary. Pauling told E. Bright Wilson Jr. that he planned to stay in Pasadena "until the last possible moment" to help the new lab workers settle in and prepare for the coming months without him. Pauling also arranged for a graduate student to work under him while at Cornell, choosing Philip Shaffer Jr. from Harvard rather than someone from Caltech. Shaffer's assistance was sufficient to merit a mention by Pauling in the preface to *The Nature of the Chemical Bond*. In the end, two Caltech research fellows also made the trip: G.C. Hampson, who continued his research on crystal structures, and H.D. Springall, who continued his work on electron diffraction. Both also earned Pauling's gratitude in the preface.

Linus and Ava Helen arrived in Ithaca during the last week of September 1937. Their date of arrival gave Linus one week to settle affairs before the start of his duties, which included delivery of the Baker Lectures on Tuesdays and Thursdays, as well as supervision of a weekly Wednesday seminar. The couple selected the Telluride House, a student residence, as their home while in Ithaca, and Pauling wound up staying there for the duration of his lectureship.

It didn't take long for Pauling to make an impact: during the third week of his visit, he gave a public lecture to an audience of 100 that drew the attention of the *Cornell Daily Sun* and the *Ithaca Journal*. The town newspaper described Pauling as building "his story around the statement that 'Structure is the basis of all chemistry.'"

At the beginning of November, once Linus was fully settled, Ava Helen returned to Pasadena. The separation was difficult for them both. By writing to each other several times a week, they salved their heartache and kept up to date on the everyday activities that occupied them and those around them. Ava Helen kept her husband informed on how Peter was beginning to read, how Linus Jr. was learning to pronounce "competitor," and how Crellin was being "such a good baby" who "literally never cries." Though Pauling missed the children, he longed for Ava Helen most of all and wrote often of his loneliness. "I love you, my own dear Ava Helen, with every bit of me," he wrote on November 20. "Life doesn't mean anything while you are away — I live in a sort of daze, with nothing worthwhile. The only thing I can stand to do is to work." Pauling wrote the bulk of *The Nature of the Chemical Bond* while at Cornell, a surge of productivity that may have been driven, at least in part, out of a motivation to suppress his longing to be with his family.

Indeed, Pauling's Cornell correspondence with his wife chronicles how hard he pushed himself to progress through his project. Writing longhand between ten and 40-plus pages per day of manuscript, Pauling often stayed late at the lab, sometimes until three or four in the morning. This upset his wife, who repeatedly admonished him for working himself too hard. On November 27 she wrote, "You are an awful boy to try to work all night. Your Wednesday (really Thursday) letter came today and I'm mad — hopping mad as Peter says. I told Mrs. Crellin that you worked until 4:10 A.M. (She took us all riding in her electric car this morning for an hour.) She said you were shortening your life and that you owed it to your family to take care of yourself. It is wonderful that you were able to get so much done but I do worry about you."

Pauling was definitely able to get a lot done, finishing more than half of the chapters for his first book draft by early December, only a month after Ava Helen had left.

*

While Pauling used his absence from family to fuel his work and writing, he also ran into a few obstacles and courted various diversions.

An early issue came by way of his left wrist, which was starting to hurt. On November 11, Pauling's main Cornell contact, Jacob Papish, intervened directly by calling a doctor "who said (without looking) for me to get it baked out at the hospital with a short-wave apparatus and bandaged. This was done."

Pauling was not too impressed by the treatment. Later in the day he reported, "I am carrying my arm in a sling — it hurts when I use it, and I find myself using it if it is free. I think it will be well soon, though. If it isn't I'll go to a doctor — not Papish's." From the sounds of it though, the treatments started to work; by the 15th, Pauling was telling his wife that "it still hurts, but only once in a while, and I can use my hand if I am careful." By the 24th, with most of his wrist pains behind him, Pauling finally remembered how he had hurt himself in the first place, recalling that he was "at Maury's office" and "fell over backward in his chair — flat on the floor — and I'm sure that I fell on my wrist. Isn't it strange that I forgot that? I'm rather tired of writing."

Pauling also found a few diversions to break up his otherwise relentless pace of writing and lecturing. Newly enmeshed in a very different academic climate, he took advantage of the opportunity to sit in on numerous campus lectures. On November 11, for example, he "listened to a long talk" that was "rather boring" but still "boasting an interesting point or two — on the sweet-potato starch industry." He also engaged in professional reading, including Edwin C. Kemble's *The Fundamental Principles of Quantum Mechanics with Elementary Applications*, as well as pleasure reading, most commonly the Sunday paper or *Time* magazine. He included a bit of space for fiction as well, including two short stories by Thomas Mann and Christopher Morley's *The Trojan Horse*, which he found "very amusing" and useful for "put[ting] me in the mood (Liny's word) for sleep." Alas, the technique didn't work too well, because the next day a weary Pauling wrote to Ava Helen that he was going home early to finish the novel and go straight to bed.

These diversions, it would seem, were not enough to slake Pauling's loneliness and he continued to seek out ways to be with Ava Helen.

At Thanksgiving Pauling wrote to his wife, "I am working hard now so that if you do come back with me in January I'll have more time to play with you. We would have fun going to Princeton and Yale (also Buffalo — we would go to Niagara Falls again). I liked having you in the lab with me, but I did get worried about you, thinking that you were bored while I was trying to work. If you come back with me I'll work in my/our room and you can read or go to bed. We used to do that in Munich. You have forgotten what it is like to have Paddy with you working."

Though Ava Helen initially protested the idea of going back to Ithaca, she gradually warmed to the suggestion. For it to happen, they needed their helper Lola Cook to take care of the four Pauling children, including Crellin, still an infant. This may not have been too difficult to arrange since Ava Helen had told her husband earlier, on November 11, that Lola "said she wants to take care of the baby!" On December 2, Ava Helen continued, "I'd leave Lola with the baby I think and get someone to do the work. I'm hoping that after three weeks at home [for Christmas] you will want to return to Ithaca alone. That would be simpler and less expensive." Those three weeks, as it turned out, were not enough, and after the holidays Ava Helen returned to upstate New York with her husband.

In the weeks prior to the Christmas holiday, Pauling found himself running out of steam. On December 3, he told Ava Helen, "I'm afraid that I'm getting stale — I've written only a few pages today." A few days later he repeated how "stale" he had become, telling his wife, "I need you to play with me and love me and make me happy again." Ava Helen responded, "Of course you can't write more on your book because you've worn yourself out." Luckily for Pauling, his plan to make a quick exit from Cornell for the winter break was successful and he was on his way home early in the month. Riding the train back to Pasadena, Pauling continued to work on his book, telling Ava Helen, "I haven't anything to read" and had, as a result, "been planning out the last chapters of the book."

Once Linus and Ava Helen were back together, first in Pasadena and later in Ithaca, progress on *The Nature of the Chemical Bond* slowed considerably — it would take several months to match the productivity of Pauling's one month alone at Cornell, during which time he had written half of his book. On February 10, 1938, Pauling, now back in Pasadena for good, wrote to Papish at Cornell to let him know that he had just received his manuscript by mail and "shall now settle down to work on it with the hope of completing it before long." Over a month later, on March 18, Pauling told his Caltech colleague Eddie Hughes, who had stayed at Cornell to help push the book through the university's press, "I haven't done very much toward completing the chemical bond book, but I hope to get to work on it soon." The following month, when Pauling was away from Ava Helen again, he told her that he was working on the text "for a while (correcting old chapters)."

By May contacts at Cornell were inquiring into the whereabouts of Pauling's book, but the author still had one chapter left to write. On May 10 he told Hughes, "I am indeed anxious to get my book finished, but I am having trouble in finding time to work on it." Pauling decided to begin sending it in sections and told Hughes that he would finish it by the end of June, at which point Hughes could make his way back to Pasadena.

Interestingly, though he had not yet begun sending the manuscript to Cornell, Pauling shifted his correspondence with Papish toward the prospect of a second edition. "[T]he field is progressing so rapidly," he wrote, "...[a second edition] probably should be prepared in about two years." A month later, Pauling finally began sending chapters one by one, telling Hughes that he was mostly finished "except for two or three odd sections" and "some of the figures." But as Pauling would find, his delay in getting a final manuscript to Cornell would cause troubles in the months to come.

*

Once the chapters had been received by the Cornell University Press, the next steps were to get the text formally approved by the Press's board of directors and to find a printer. As it turned out, these tasks were not simple ones to achieve. First, Pauling needed to deliver a complete manuscript that could be approved by the board. To do this he leaned on his Caltech colleague Eddie Hughes, still residing in Ithaca, who served as an intermediary with the Press and made several last-minute changes to the manuscript, as directed by Pauling from Pasadena.

Hughes was finally able to hand over a finished product to editor W.S. Schaefer on July 5, 1938. A quick turnaround to print looked dicey however as, according to Hughes, "it was the seventh book they've had ready for press in the past three weeks." Schaefer promised Hughes that "the typescript would be in the printer's hands at least before August 15," but "it will be impossible to have the book before November 1 at the very earliest." This timeline was further interrupted in late July when Schaefer broke his knee and needed to be hospitalized. Hughes, for his part, had done all that he could and made his way back to Pasadena at the end of August.

It didn't take long for Pauling to grow anxious. On September 9 he wrote to both Schaefer and Papish, asking for any word on the book's progress, and noting his desire to use it in one of his upcoming courses. Schaefer responded apologetically, informing Pauling about his knee and how "the manuscript was received too late for publication this fall" because, over the summer when the book had arrived, "it was no longer possible to assemble our Committee on Publication." The committee was scheduled to meet within the next week, and Schaefer assured Pauling that he would "rush" publication as much as possible.

Pauling did not take this news well, and immediately wrote back to Schaefer that he had only just heard of this delay due "to the failure of your Committee on Publication." Pauling was even more forthright with Papish, blaming Schaefer for "still holding up the printing until this Committee meets" and calling the projected two-and-a-half-month delay "inexcusable." Overlooking his own earlier slowness in getting the manuscript compiled, Pauling wrote, "if I had known at the beginning of the summer that this delay was contemplated, I might have done something about it."

Papish tried to direct Pauling's ire away from Schaefer, telling him "the [delay] was not due to inefficiency or procrastination on the part of Mr. Schaefer but to the organization of the University Press" which, being composed of professors, "is not very active during the summer." This explanation in hand, Pauling seemed to have accepted the book's fate, as he adjusted his own course schedule to incorporate the backed-up publication. He also began explaining to others that his book might not be out until January, a date that, as it turned out, was overly optimistic.

With the book in press by the first weeks of November 1938, Pauling began to send in revisions to Schaefer, arguing that "the unexpected delay of three months has caused the book to be somewhat out of date in places." Schaefer was amenable to including these changes, telling Pauling that doing so would delay the galley proofs somewhat, but that this loss "will be made up later." Pauling started receiving, correcting, and returning the galleys to Schaefer by the end of December, and after getting through the first four chapters, Schaefer suggested that Pauling send the remainder directly to the printer in Wisconsin, in order to speed things up.

In January 1939, with the process moving along, Pauling began receiving more and more inquiries on the status of his book. Pauling had to tell retailers and professors alike that they would all have to wait — the initial word was that it would be ready in March, which quickly became April. During this time, Pauling also decided to add a 12th chapter, (titled "A Summarizing Discussion of Resonance and its Significance for Chemistry") and suggested to Schaefer that he prepare it for print without the customary galley review. Pauling still hoped to have the book ready for at least part of his spring course.

Since all of the galleys had been returned by the end of January, Pauling was expecting the page proofs to arrive by the end of February. When this did not come to pass, Pauling became agitated again, reminding Schaefer that he was "being inconvenienced this year in teaching my class here because of lack of the book." The page proofs started to arrive at the end of March and order arrangements were made for the Caltech bookstore, but interferences kept arising — now an influenza outbreak had rendered the printer short-handed. Schaefer told Pauling that outbreak "they are doing their utmost to complete your book," but they did not want to "risk using inexperienced men for this difficult task." As Schaefer's letter was en route, Pauling, responding to an earlier request to shorten his preface, asked the Press to add Eddie Hughes to the list of individuals thanked for helping him put the book together. In this same letter, Pauling asked Schaefer to remove the printer from those so acknowledged.

In mid-April, Pauling returned the final proofs, correcting any left-over spelling mistakes, and soon followed up with author and subject indices. It was at this point that Pauling also sought to negotiate the price of the book. Forgetting that he had earlier expected it to sell for one cent per page, Pauling suggested that the 430-page book be priced at three dollars, and no more than three dollars and fifty cents. Schaefer told Pauling that this price was not possible, partly due to the special mathematical and chemical notations required by the book. Additionally, previous books in the Baker Lecture series had been published through the Chemistry department at Cornell which had "rather lush" funding. As the series was now fully under the purview of the University Press, the selling price would need to cover production costs. Schaefer suggested a list price of four dollars and fifty cents, knowing that the Press would not garner this full amount on most sales because of discounts afforded to educational organizations, booksellers, and foreign distributors. Pauling accepted Schaefer's price point; mostly he was anxious to get the book printed.

On May 8, Schaefer wrote to Pauling that, at last, the day had come: *The Nature of the Chemical Bond* was finally being printed, nearly six months after the original date proposed. Nonetheless, the books that Pauling ordered for his students did not reach Caltech until May 29. Pauling lamented to Schaefer that this "was during the last of my lectures for the year, so that it turned out that the students were not able to use it very much in connection with my course."

*

Once all of the hindrances to getting *The Nature of the Chemical Bond* printed had finally been overcome, Pauling began work on the project's next phase: promotion. He started by compiling a list of people to whom he wanted to send the text, a roster that included those who had helped him along the way, journals that would review the book, previous Baker Lecturers, and chemistry professors who might be interested in using it in their courses. The final list consisted of 61 names, not counting journals.

Upon receiving the names, Schaefer replied that the press's policy was to allow for only six free copies, but that he could send the book to all those identified at a 33% discount. Pauling explained that he had drafted the list based on previous experiences with McGraw-Hill, which was much looser in doling out free copies. He then asked Schaefer to remove 16 individuals from the list, paid for four copies, and suggested that the remaining individuals were likely to use it in the classroom and should receive a copy as a promotional offer. Schaefer agreed to these terms, charging the bulk of the books to the press's advertising budget.

Once *The Nature of the Chemical Bond* was officially released in May 1939, Cornell did its part to spread the word. For starters, the press sent out an order form addressed to "Students of Chemistry and Molecular Structure" in chemistry departments across the country, alerting them to the opportunity of a 10% educational discount. The form summarized the thrusts of the book as including "the structure of molecules and crystals, and the nature of the chemical bond," and emphasized its grounding in quantum mechanics without relying on "mathematical argument in demonstrating the conclusions reached." The flyer continued:

> "Early chapters discuss the theoretical basis and the nature and proper-
> ties of isolated bonds between pairs of atoms. Then complex ions, mol-
> ecules, and crystals are considered; the extensive illustrative material is
> drawn about equally from organic and inorganic chemistry. There are
> complete chapters on such important subjects as the hydrogen bond,

ionic crystals, and metals. The methods used by the author in exploring the nature of the chemical bond include the resonance concept and the techniques of diffraction of electrons by gases and vapors and of X-rays by crystals, the determination of electric and magnetic moments, and various kinds of thermal measurements."

The press also produced an advertising brochure that expanded upon the information contained in the order form. In reiterating the book's grounding in quantum mechanics, the brochure promised an "especial emphasis on the resonance phenomenon, including the new concept of resonance of molecules among alternative electronic structures." It also positioned the book as being "of unusual importance for chemists and mineralogists," and quoted from a glowing write-up in the August 1939 *Scientific Book Club Review*, which declared that "the publication of this book is literally epoch making." The journal also described how Pauling's final chapter, which he added at the very last minute, dealt "with the future development and application of the concept of resonance" and "will probably prove to have been truly prophetic."

By June and July, readers began corresponding with Pauling about the book. The initial responses, mostly from academics, thanked Pauling for sending a complimentary copy and were generally positive in their evaluation. Many also mentioned how they, or someone in their department, planned to use the text in an upcoming course. Earl C. Gilbert of Oregon State College, for example, told Pauling that the volume would "be very successful and fill quite a need." Gilbert singled out Pauling's account of hydrogen bond properties as a particularly "convincing treatment" in comparison to Jack Sherman and J.A.A. Ketelaar's application of quantum mechanics to the carbon-chlorine bond.

Joseph E. Mayer of Columbia University was more effusive in his praise, writing, "It's the first book that I've read through for years!" C.P. Smyth issued a similar response, telling Pauling in mid-October, "As evidence of my interest in it I can cite the fact that it is the first scientific book which I can remember reading during the course of a fishing trip, although I have carried many with me in the past." Pauling also received encouragement from a former mentor, G.N. Lewis at Berkeley, who wrote in August, "I have returned from a short vacation for which the only books I took were a half a dozen detective stories and your 'Chemical Bond.' I found yours the most exciting of the lot."

Pauling appreciated the responses and was particularly glad that Lewis was happy, as he had dedicated the book to him. In his response, Pauling confided "that I had you in mind continually while it was being written, and I have been hoping that my treatment would prove acceptable to you."

Along with the praise, Pauling also received constructive advice, which he was eager to incorporate into a second edition. Joseph Mayer mentioned that the book's discussion of metals needed some work, and in a letter to Gerold Schwarzenbach of the University of Zurich, Pauling noted that he "hoped to give proper discussion" of Schwarzenbach's findings on acid strengths "in a revised edition..."

Buoyed by the feedback he was receiving, by mid-October Pauling began to suspect that he had a hit on his hands. Curious about sales numbers, he wrote to Schaefer for an update and also asked for an interleaved copy that he could use to plan out the next edition, which now seemed a foregone conclusion.

*

In addition to responses by academics, Pauling received plenty of reaction from students. Lois Joyce was one such correspondent. Working her way through graduate school at the University of Illinois, she began contacting Pauling in May 1939 in hopes that she could study with him at Caltech, telling him that she was much more interested in his focus on molecular structure than on the analytical chemistry that she was studying at Illinois. After bringing her case before the Division of Chemistry and Chemical Engineering at Caltech, Pauling was compelled to tell Joyce that he could not bring her aboard due to Caltech's restriction on women without PhDs working in their labs. But he encouraged Joyce to continue on and get her doctorate at the University of Illinois, after which point she might be able to join him.

A month later, Pauling received a letter from Joyce's mother asking for an autographed copy of what she could only remember as "The Strength of the Chemical Bond," to be presented to her daughter on her birthday. She told Pauling how Joyce regarded him as "one of the greatest men in the world" and that even though "it's a strange career for a girl" her daughter was "deeply interested in X-ray research and willing to give up all pleasure in life to succeed," often studying "until two and three in the morning." Pauling obliged by signing and sending a copy of his book through special delivery, as Joyce's birthday was only days away. The recipient was overjoyed when she received the gift, telling Pauling of her gratitude that her mother had "bothered" him to send a copy and that she was "never so thrilled." Joyce still regretted that she was unable to work with Pauling however, telling him again that the only topic she wanted to study was molecular structure.

Scholarly reviews of *The Nature of the Chemical Bond* did not start appearing until 1940, the year following its publication. Many of the reviews offered soaring praise not only for the book, but for Pauling as well. They also included more technical criticism than had been contained in the letters that Pauling had received earlier from his colleagues.

The *Transactions of the Faraday Society* published a review, authored by one L. E. S., which proffered that "All who have researched in the field of molecular structure have long awaited this book." According to the review, Pauling's account mostly focused on "the structure of individual molecules...with problems centering around relatively normal covalent bonds." It also included a "delightful chapter" on the hydrogen bond and two "succinctly but excellently discussed" chapters on the structure of crystals.

Similar praise came from John E. Vance in the *American Journal of Science* and George B. Kistiakowsky in the *Journal of the American Chemical Society*, both of whom lauded Pauling's non-mathematical style. Kistiakowsky added that the presentation was "by and large...lucid, and a student with little more preparation than the four basic courses in chemistry should be able to digest most of the contents." Indeed, even novices might find the text "stimulating." The reviews from Germany echoed these sentiments, though one author in the *Zeitschrift des Vereins Deutscher Chemiker* was disappointed that Pauling, like most English chemists, glossed over the German literature on the subject.

As the reviewers continued, they brought in a handful of criticisms of Pauling's work. L. E. S. found that, in places, Pauling had oversimplified his discussion and concluded with a "premature happy ending." According to the author, these flaws emerged from Pauling having written such a broad survey and resulted in assumptions along the lines of "the electric dipole moment of a purely covalent link is small or is zero," a suggestion presented without any proof. L. E. S. believed that Pauling's certainty in his own understanding came "partly from this simplification and partly from tricks of style."

Kistiakowsky put forth a similar observation, calling attention to Pauling's "pontifical style" and "his advocacy of the doctrine of infallibility of Pasadenean research," before offering that this approach was "understandable and should not be taken amiss."

A more serious criticism came from Vance, writing in December 1939, who complained that Pauling ignored a large stream of the chemical discourse then current. In particular, Vance found it "remarkable that no mention of the Hund-Mulliken treatment of chemical bonds," based on molecular orbitals, appeared in the text. Such an inclusion, Vance thought, surely would have increased the "usefulness of the book." He was not the only person who felt this way.

In June 1940, Robert S. Mulliken — himself a rising scientific star and future Nobel Prize winner whose atomic structure work was sometimes at odds with Pauling's — published his own review in the *Journal of Physical Chemistry*. Mulliken initially adopted a generous tone, calling Pauling's book a "clearly written survey of the nature of the chemical bond," before adding the critical clause, "from the viewpoint of the atomic orbital method." This viewpoint, Mulliken ceded, was most suited to combining wave mechanics with the "traditional ideas" of chemical bonds and therefore had "wide appeal and usefulness among chemists." But it was not the only perspective in play. Indeed, Mulliken's molecular orbital model, developed with Friedrich Hund, was a competing body of work that Pauling had largely ignored.

Mulliken suggested that Pauling's failure to include a molecular orbital perspective, except for a brief "aside," was misleading, especially for the "unfamiliar reader." "Most authorities," according to Mulliken, "would feel that for a deeper understanding of the electronic structures of molecules a knowledge of both methods is necessary, and that for many problems the MO [molecular orbital] method is the simpler and more intelligible." Despite this criticism, which became more common over the years, Mulliken ended on an upbeat note, echoing his earlier comment about the volume's usefulness and confirming that "the book is a landmark in the history of valence theory."

While criticism urging Pauling to add another theoretical perspective to *The Nature of the Chemical Bond* may not have been exactly what he was looking for, Pauling was not able to incorporate any of the new feedback into the next edition anyway. After he had received his requested interleaved copy from W.S. Schaefer back in October 1939, Pauling quickly got to work on making revisions. By December he had already sent the first three chapters to Cornell, and by May 1940 the second edition was out, published one year after the first edition and one month prior to Mulliken's review, also of the first edition. The opportunity to incorporate new suggestions was not completely lost however, as Pauling began working on yet another version of his landmark text just one year later, in 1941.

Pauling's Battle with Glomerulonephritis

Chapter 2

By Carly Dougher

Pauling family portrait taken in 1941. The back of this photograph is annotated, "1941. Daddy very ill."

On March 7, 1941, Linus Pauling stood before a gathering of colleagues, prepared to deliver an acceptance address for the prestigious William H. Nichols Gold Medal, presented by the New York Chapter of the American Chemical Society. Before beginning his remarks, Pauling spoke candidly to his audience. He thanked the award committee for his selection and expressed gratitude that the ceremony had provided him with an opportunity to reconnect with old friends.

On this occasion however, it was apparent to all in attendance that Pauling's physical health was suffering. His face was bloated and he reportedly lacked the enthusiasm for which he was so well-known. Addressing the elephant in the room, Pauling joked, "Several of [my old friends] said to me tonight that I appeared to be getting fat. This is not so."

In fact, just that morning, Pauling had awoken to find his face so inflamed that his eyes were nearly swollen shut. Over the previous few weeks, he had been experiencing noticeable swelling, weight gain, and chronic fatigue but he could not identify the cause of his ailments. On the date of his talk, his tongue felt enlarged and his voice was flat.

With his audience, Pauling half-heartedly pondered the cause of his puffed-up appearance and jabbed at his illness for laughs. In comparing the experience to childhood encounters with poison oak, Pauling joked that "Yesterday I must have bumped into something similar," and continued that "...while I was wondering what the responsible protein could have been, I decided that it was a visitation — that I was being punished for thinking wicked thoughts."

The following evening, Linus and Ava Helen had dinner at biologist Alfred Mirsky's residence. While there, Pauling was examined by another guest, Dr. Alfred E. Cohen, a cardiac specialist from the Rockefeller Medical Institute. After ruling out problems with Pauling's heart, Cohen expressed puzzlement at Pauling's condition. Nothing appeared to be wrong with the 40-year-old man, other than his extreme edema. But Cohen was concerned by the severity of the swelling and recommended that Pauling come into his office the following day for a more thorough examination and lab work-up.

Adhering to the physician's advice, the Paulings met with Cohen in his office at the Rockefeller Medical Institute, only to leave with alarming news. After a battery of lab tests had been completed, Pauling was diagnosed with Bright's disease — a serious renal disease that results in the degradation of the kidneys. At the time, little was known about the condition and the majority of the medical community considered it to be a terminal diagnosis.

From there, Pauling was fortunately referred to a leading special-ist, Dr. Thomas Addis, who was the head of the Clinic for Renal Disease at Stanford University. Addis was a pioneer in the field of nephrology and his treatment plan, at the time, was both novel and revolution-ary. Had Pauling not been referred to Addis's care, the treatment he would have received elsewhere — polysaccharide infusions to reduce his edema — would almost surely have resulted in his early death. But under the guidance of Thomas Addis, Pauling's condition was effec-tively treated through alternative means. By May, Pauling was reporting improvements in his overall well-being and, by August, the edema had completely disappeared.

Since Pauling's time of diagnosis, Bright's disease has been reclas-sified and redefined. Now it is believed that Pauling was suffering from what is currently termed acute glomerulonephritis, a condition that is characterized by inflammation of the kidneys caused by an immu-nological response. Damage to the small clusters of capillaries within the kidney, known as glomeruli, results in what might be simplistically described as a "leaky kidney." When the glomeruli are damaged, pro-teins leak from the bloodstream into the urine through the damaged portions of the kidney. Thus glomerulonephritis consequentially leads to excessive protein loss. As it worsens, the condition profoundly impacts the body's ability to function, because the nephritic kidneys are unable to properly filter the blood.

In his 1941 speech, Pauling had wondered aloud about a protein that might be responsible for his swollen condition. That culprit protein was probably albumin. As proteins leak from the bloodstream into the urine, blood proteins, called albumin, exit the bloodstream. These pro-teins are known to be essential to the regulation of blood osmotic pres-sure, and without sufficient albumin in the bloodstream, the affected system becomes incapable of efficiently extracting excess fluid from the

body cavity. This excess fluid then remains trapped in the body and ultimately results in swelling along the lines of what Pauling was experiencing in 1941.

Although the albumin did not cause Pauling's condition, the loss of
this blood protein due to the nephritis appears to have resulted in the
symptoms that he was experiencing at his award ceremony. Therefore,
contrary to his original speculation, it was the absence, rather than the
presence, of a protein that caused his extreme fluid retention.

*

Thomas Addis, known to close friends and colleagues as Tom, was born
on July 27, 1881 in Edinburgh, Scotland. After completing a stint as a
Carnegie Research Scholar and Fellow, Addis left Europe in 1911 to pursue a career as a clinical investigator at Stanford Medical School. Upon
arriving in the United States, Addis devoted his life's work to the study of
kidney disease. By the time of his death at age 67, Addis's achievements
were immense and widely recognized. Quoting William Dock, an MD
at the State University of New York, "As a medical scientist [Addis] was
in a class by himself."

During his lifetime, Addis published more than 130 scientific papers
as well as two books on renal disease, *The Renal Lesion in Bright's Disease* (1931) and *Glomerular Nephritis: Diagnosis and Treatment* (1948),
both of which were well-received in the medical community. As Stanford University medical professor Arthur Leonard Bloomfield would
recount, "Addis's book with Oliver on the renal lesion in Bright's disease
is, of course, a classic, but the little volume on glomerular nephritis completed only a few months before his death seems to embody his philosophy of disease and of science in general; it will perhaps interpret the man
to his followers better than anything else he has done."

Known for his unique laboratory structure, Addis strongly encouraged collaboration between all members of his scientific team, and
managed his operations based on what he believed to be "democratic
centralist principles." (Among his coworkers was his wife Elesa, a lab
dietician.) Throughout his career, Addis's loyalty to the group never

wavered. Pele Edises, a writer for *People's World*, once requested a profile interview with Addis, to which the physician replied, "Why can't you just write about the lab and leave me out of it?"

As Pauling's treatment moved forward, he and Addis became both scientific colleagues and good friends. Importantly, beyond their shared professional interests, Pauling and Addis also maintained characteristically active political minds. Outspoken in his beliefs, Addis's political leanings led him to create and chair the San Francisco Chapter of the Spanish Refugee Appeal. A primary activity of the Appeal was to raise funds for a clinic in Toulouse, France, known as the Varsovie Hospital, which was dedicated to the care and rehabilitation of Spanish Republican refugees.

Like Pauling, Addis's political inclinations met with significant resistance throughout his professional career. In a typescript dictated for Addis's memorial, Pauling noted that his doctor's affiliation with the American Medical Association (AMA) had been turbulent throughout his lifetime due to his politics. In one instance, Addis raised objections to the California Medical Association's support of a coffee cancer cure, which Addis believed to be exploitative of individuals seeking treatment. On another occasion, Addis spoke out against a $25 contribution requested or required by the AMA to fight against President Truman's system of medical insurance.

Addis passed away on June 4, 1949, at Cedars of Lebanon Hospital in Los Angeles. His death was mourned both at home and abroad, with the October 1949 newsletter of the Varsovie Hospital reporting:

> *Con la muerte del Dr. Thomas Addis, los antifranquistas, los republicanos españoles, perdemos un gran amigo y un valiente luchador en defensa de nuestra causa, por la República, por la Democracia y por la Paz.* ["With the death of Dr. Thomas Addis, the anti-fascist Spanish Republicans lost a great friend and a valiant fighter in the defense of our cause, for the Republic, for Democracy, and for peace."]

In his honor, a new laboratory was added to the outpatient clinic of the Varsovie Hospital in 1950. And long after his death, Addis's name was used effectively to raise funds for the clinic and to provide aid to Spanish refugees victimized by the Franco regime.

In 1955 Linus Pauling was awarded the first Thomas Addis Memorial Award by the Los Angeles Chapter of the National Nephrosis Foundation. The award would evolve into an annual honor granted in recognition of significant contributions to the study of the kidney and its diseases.

After Addis's death, Pauling offered to write a memoriam detailing Addis's life and accomplishments. In response, the Addis family asked that no mention of his political affiliations be included, for fear of their own personal safety. Wishing to honor this request, Pauling and his co-author, Dr. Richard Lippman, made a decision to delay publication of the piece. In Lippman's estimation, "It is impossible to characterize Dr. Addis...without some discussion of his political ideas and his philosophy of politics and people." A revised version of the piece, co-authored by Pauling and Kevin Lemley, would not appear in print until 1994, the year of Pauling's death.

*

By studying the urinary contents of both healthy individuals as well as individuals affected by Bright's disease, Addis was able to develop a systematic treatment plan unlike any other of the time. And as his methods refined, Addis was increasingly able to provide what most other kidney specialists at the time could not: hope and options.

Throughout his career, Addis consciously integrated laboratory research into his clinical practice by methodically "mixing patients with rats." In his effort to thoroughly understand the kidney and its diseases, Addis followed a group of nephritics for over 30 years while also maintaining a colony of more than 1,000 rodents for *in vivo* experimentation. So entwined were the two that Addis increasingly refused to separate his clinical work from his laboratory studies, going so far as to see his patients in a small cubical in the corner of the laboratory.

From his lab experiments and clinical experiences, Addis developed a unique method for the treatment of nephritis. Based on the physical concepts of work and healing, Addis believed that in order for the kidneys to recover, the patient must allow the organs time to rest. In pursuit of a reduced renal workload, Addis prescribed a low-protein, low-sodium diet for his nephritic patients. This method appears to have worked effectively for the treatment of many (but not all) of Addis's patients and proved to be the solution for Pauling's ailments.

Addis's treatment system specifically relied upon a series of quantitative measurements that soon became known as the "Addis urea ratio" and the "Addis count." The aim of the "Addis count" was to determine the number of red blood cells, white blood cells and casts (clumps of red and white blood cells) excreted in the urine, while the "Addis urea ratio" measured the concentration of urea in the urine. Addis used these two data points, in addition to a measurement of total urine output, to assess the nature and extent of a patient's disease and to track the progress of their treatment.

When Pauling came to Addis in 1941, he was quickly impressed by the physician's methodology. In addition to the measurements described above, Addis qualitatively examined the cells obtained from Pauling's urine sample and identified any deformations and abnormalities that he observed.

Although Addis relied heavily on his laboratory group, he conducted all of his patients' sample analyses himself, arguing that in order for the data to have meaning, the physician must be responsible for all work of this sort. After gathering sufficient information, Addis used his qualitative and quantitative findings, as well as his experience with his patients, to clinically reclassify Bright's disease and develop tailor-made treatment plans.

While Addis's treatment offered relief to many, his findings and conclusions were often criticized by the scientific community for lacking validity. These critiques stemmed primarily from Addis's refusal to use controls in his clinical studies. Addis firmly believed that controls in his research would be unethical. His duty, he wrote, was "to treat each patient in the way that I think will do him the most good." Were Addis to have incorporated controls into his clinical research, he would have been forced to split his patients into two groups. This experimental design would have left half of his patients without the treatment that he, as their physician, believed to be most effective. Addis considered an approach of this sort to be unacceptable. As a result, the majority of the scientific community continued to cast a skeptical eye towards Addis's contributions.

*

While Addis is rightfully credited with Pauling's recovery from nephritis, Addis himself preferred to place the acclaim elsewhere. Over the course of his practice, Addis relied heavily on the support of his patients' families and was convinced that their efforts were vital to his patients' care. It is the "...wives, mothers, and sisters," he once said, "who, with our patients, are our true colleagues with whom we work and for whom we work."

Ava Helen Pauling was clearly no exception and it did not take Addis long to acknowledge her as his colleague. Throughout Linus's recovery, Ava Helen and Addis sent letters back and forth updating one another on his status, with Addis commonly signing his letters as "Your ever faithful collaborator, T. Addis." For her part, Ava Helen often referred to Pauling as "our patient" in her responses back.

As soon as Pauling started the low-protein, low-sodium diet that Addis had prescribed, Ava Helen began keeping record of her husband's intake. Her notebook is meticulous and thorough, and occasionally includes small notes about Pauling's improving health and daily activities.

The log of Pauling's diet begins on April 9, 1941. In each of her entries, Ava Helen documented what and how much her husband ate, as well as the protein content (in grams), the salt content (in milligrams), and the calories in all that he consumed. She tallied the amounts after each meal and recorded final totals at the end of the day. With the support of his loving and dedicated wife, Pauling was able to stay on this strict diet for 15 years.

Each day during those years, Pauling consumed between 2,000–3,000 calories (usually closer to 3,000), around 55 grams of protein, and approximately 1.2–1.6 grams of salt. His breakfasts typically consisted of a citrus fruit or glass of fruit juice, a cereal (shredded wheat, pancakes, or crepes suzette), milk, cream, and a cup of coffee. For lunch he generally ate various combinations of eggs, water biscuits, apple sauce, potatoes, cheese, fruits, and vegetables, often followed by a chocolate bar. (From April 9 to June 30, 1941, Pauling consumed 65 chocolate bars.) In the evenings he would dine on fruits and vegetables, cheese, water biscuits, baked potatoes, and milk.

Below is an example of a typical day of Pauling's nephritis diet.

Wednesday, April 23, 1941

Breakfast
- 1/2 Grapefruit
- 6 Pancakes
- 2 Squares of Butter
- 6 Tbs Syrup
- 2 Tbs Cream
- 1 Cup of Coffee

Breakfast Total: Protein 11 g; Calories 1,050; Salt 259 mg

Lunch
- 1 1/2 Cup Eggnog
- 1 Piece of Coconut
- 1 Medium Orange
- 1 Medium Pear
- 1 Chocolate Bar

Lunch Total: Protein 20 g; Calories 858; Salt 543 mg

Dinner
- 1 Baked Potato
- 1 Square Butter
- 2/3 Cup of Cabbage
- 1 Cup of Milk
- 1/2 Cup of Gelatin

- 1/4 Cup Cream (30%)
- 4 Cookies

Dinner Total: Protein 19 g; Calories 830; Salt 510 mg

Daily Total: Protein 54 g; Calories 3,020; Salt 1,312 mg

It is apparent that Pauling had a sweet tooth as his dietary record is littered with notes on sugary treats such as custards with chocolate sauce, sponge cakes, meringues, ice cream, fruit tarts and pies, puddings, cookies, Coca Cola, and strawberries with cream. But in this instance it was the protein and salt that truly mattered, and before long Pauling was on the road to recovery.

*

Currently in Western medicine, it is believed that nephritis is caused by a variety of health conditions ranging from acute infections to autoimmune disorders. The standard of contemporary care includes corticosteroids, immunosuppressant drugs, antibiotics, plasmapheresis, renal dialysis, and kidney transplants. Corticosteroids are used to reduce inflammation of the kidneys and immunosuppressant drugs are prescribed to limit the immune response that triggers the inflammation. If the cause of an individual's nephritis is known to be bacterial, antibiotics are administered. In severe cases that progress rapidly to kidney failure or end-stage kidney disease, plasmapheresis, renal dialysis, and kidney transplants are prescribed.

The exact cause of Linus Pauling's nephritis remains unknown. And while the low-sodium, low-protein diet was clearly successful in returning Pauling to health, the dietary therapy for nephritis lost popularity within the United States shortly after Addis died. Instead, Addis's protocol was largely replaced by steroid therapies and renal dialysis.

The medical community seems to have mostly lost interest in Addis's dietary therapy until a 1982 *New England Journal of Medicine* article that reconsidered the treatment. The primary author of the paper, Barry Brenner, extended Addis's thinking, providing additional insight into the progression of damage to the glomeruli caused by nephritis.

Using rats as a model, Brenner determined that, upon the onset of damage to the glomeruli (the small capillaries), all remaining healthy glomeruli begin overcompensating for the lost filtration function. By elevating internal pressure, filtration through the remaining functioning glomeruli increases. This overcompensation presents severe consequences for the healthy glomeruli. As they continue to compensate, the healthy glomeruli bear increased stress, causing wear-and-tear on their structure that ultimately results in additional damage to the remaining functioning glomeruli. Brenner termed this form of compensation "hyperfiltration." In revisiting Addis's proposed low-protein remedy, Brenner and his team determined that a protein-restricted diet would in fact reduce the kidneys' overall workload, and thus decrease the stress caused by hyperfiltration.

Brenner's work was well-received and current treatment plans, once again, often include dietary restriction of protein, sodium, and potassium, in addition to the other aforementioned modern therapies. Whether commonly recognized or not, Thomas Addis's pioneering research on diseases of the kidney — research which, in the early 1940s, saved Linus Pauling's life — is still making an impact today.

W.H. Freeman & Co.

By Dani Tellvik

THE FOUNDATION of our publishing house is laid and we have arrived at the most pleasing experience of our young career . . . that of announcing the publication of our first book

Linus Pauling

General Chemistry

An Introduction to Descriptive Chemistry
and Modern Chemical Theory
by

Linus Pauling

LINUS PAULING has a driving curiosity about chemistry that has urged him on to a pursuit of the subject, extending in one direction into experimental and theoretical physics especially as applied to organic chemistry, and in the other direction, in recent years, into biology and medical research. Since 1938, Dr. Pauling has been in charge of the introductory course in chemistry at the California Institute of Technology. Dr. Pauling will occupy, in 1948, the Eastman Chair as Professor of Chemistry, at Oxford University, Oxford, England.

In *General Chemistry*, Linus Pauling has written a book to meet his conception of what an introductory chemistry text should be, namely a clearly presented treatment of the subject as a unified theory rather than a grouping of isolated facts.

Here, chemistry is projected as a vital subject consonant with the knowledge of chemistry that we have today. The lucidity of the text, the sweep of its sequence, the illustrations by Roger Hayward that act so well to supplement the meaning of the text — all these contribute to the merit of the book from the student's point of view. We feel, those of us who have worked on *General Chemistry*, that the author brings to the study of the subject a significance that gives to an introductory course wide scope and meaning.

Examination copies available in May
576 pp. 150 illustrations. Probable price, $4.25

W. H. Freeman and Company

549 Market Street, San Francisco 5, California

William Hazen Freeman Jr. was born in New York state in March 1905. Though information about his early life is scant, we do know that his father, William Hazen Freeman, was a doctor who specialized in gastrointestinal issues. We also know that the younger Freeman attended Hamilton College in New York and graduated in 1926, a member of the same class as famed behaviorist B.F. Skinner. Macmillan Publishing Company had already hired Freeman by the time of his graduation and, over the next few years, he rose to an editorial position within the company's textbook department. In the midst of this, Freeman relocated to San Francisco, where he began work at a newly opened satellite branch.

Not long after arriving on the West Coast, Freeman met with Linus Pauling for the first time. Macmillan was keen to publish a series of textbooks in chemistry, and Freeman felt that Pauling would be the ideal editor for such a series. Freeman also viewed any potential partnership as being mutually beneficial: Macmillan would enlist the skill, expertise, and reputation of a prominent scientific figure who had built a strong reputation as a teacher, and Pauling could use the association with a large publishing firm to develop his own series of textbooks.

Though Pauling couldn't deny the appeal of this possibility, he hesitated once he learned more details. For one, he disliked the idea that he wouldn't retain authority over the direction of the series. Rather, Macmillan planned to select the books that would be published and then pass them on to Pauling for his input. Moreover, the publisher wanted Pauling to coordinate closely with a team of Macmillan editors throughout the process. The company also intimated that they were likely to terminate the series after just a few years. Taking all of these factors into consideration, Pauling thanked Freeman for his time and declined the offer. In 1941, when Pauling began circulating early drafts of his textbook, *General Chemistry*, Freeman approached him again, this time to express Macmillan's interest in the manuscript. Though Pauling chose not to publish with the company, the two stayed in contact through the war years, a time period during which Pauling opted to postpone work on all major writing projects.

While Freeman was working in San Francisco, he met Verne Kopplin, a young lawyer who was specializing in tax law. A graduate of the University of Wisconsin, Kopplin was the 106th woman to practice law in the state. Having passed the bar in both Massachusetts and California, she was ultimately hired by the prestigious San Francisco firm of Rogers and Clark, which had previously abstained from employing women as attorneys. By the time she became romantically involved with Freeman, Kopplin had already challenged discriminatory gender practices in three states.

Freeman and Kopplin married in 1946, the same year that W.H. Freeman & Company opened for business. In each other, they recognized a mutual determination to succeed, even if they had to challenge powerful institutions to do so. For Freeman, this ambition meant leaving Macmillan when it failed to show enough genuine interest in Pauling's *General Chemistry* manuscript. Freeman also fiercely believed in small, independent presses and the importance of developing a trusting and intimate relationship between author and publisher. He once remarked to illustrator Roger Hayward that "the relationship between an author and a publisher is something like a marriage."

Macmillan's lackluster interest in Pauling's text was indeed the spark that led Freeman to create his own publishing house, and it was a gamble that paid off. In 1947, W.H. Freeman & Co. published its first book, *General Chemistry*, now regarded to be a classic of the genre. That same year also brought success for Verne, as it marked the beginning of her tenure at Rogers and Clark. She remained there for eight years before establishing her own practice. During her time in California, she also acted as a consultant for Freeman & Co., typically providing guidance on matters related to stockholders.

Freeman set up shop on Market Street, a major San Francisco thoroughfare that promised high visibility for his fledgling company. The boldness of Freeman's vision matched well with the prominence of his chosen location. As he stated in his first corporate mission statement: "We say first of all that we want each book we publish to be something that hasn't been done or that has not been well done." As an initial order of business in pursuing this objective, he revived the idea of a chemistry series for which Pauling would be the sole editor.

Once Pauling was officially on board, Freeman began sending his new editor several manuscripts per month, eager for Pauling's feedback about each submission's potential for success. In addition to providing suggestions on whether or not a given book should be published, Pauling's responsibilities as editor of the series also included bringing new manuscripts to Freeman's attention. Once a proposal was accepted, Pauling provided further suggestions as the text went through the process of development. He did all of this while also writing and editing his own books.

Pauling was an exacting reader and often much sterner than Freeman in his evaluations, rejecting most of the manuscripts that Freeman and, later, Stanley Schaefer sent to him. For one physical chemistry proposal that Pauling judged to be particularly poor, he told Freeman, "If you have any thought of publishing this, send it back to me and I'll tear it to pieces for you." Freeman replied sympathetically, noting that he found life as an editor to be mostly unsatisfactory, except in cases where he encountered writing like Pauling's.

Those few manuscripts that did receive Pauling's blessing were generally qualified with a long list of criticisms. For a general chemistry treatment authored by Arthur Campbell, Pauling sent four to five pages of revisions per single manuscript page (the book took several years to publish). But it wasn't Pauling's intention to be unduly harsh. Rather, his thorough critiques reflected his steadfast commitment to improving the quality of scientific education. Fundamentally, it was important to Pauling that his series only release texts that were current, unique, and highly effective. In this, he did much to reinforce Freeman's mission statement.

However, as the 1950s advanced and Pauling became busier, he found that there was far less time to devote to editorial responsibilities. By 1956 Freeman had become fairly concerned and, in one instance, he pressed Pauling to take a look at a book on hydrogen bonding. In doing so, he reiterated that, "I am very anxious to publish something worthwhile and of a specialized nature in your series, for it is not good to have your series show as little activity as it has." Thus motivated, Pauling refocused his efforts and returned the manuscript a short while later, offering few critical comments.

Though the series was an overall success, not every project worked out. One particularly tantalizing idea that Freeman put forth was for Pauling to collaborate with Caltech colleagues Richard Badger, David Shoemaker, and Stuart Bates to compile and contextualize a series of notes written by Roscoe Dickinson. (Pauling's major professor and the first person to earn a PhD from Caltech, Dickinson died in 1945 at the young age of 51.) Freeman's idea came from requests that a book be written on Dickinson's unpublished work in chemical thermodynamics. Unfortunately, the project never came to fruition for a number of reasons, including lack of access to Dickinson's papers.

Another possibility that went unrealized was one that Pauling flirted with from time to time: writing a high school textbook. Freeman wasn't keen on this idea, and successfully dissuaded Pauling from pursuing it by enlisting the help of his personal secretary, Margaret Cooper. Freeman had hired Cooper in 1953 and quickly came to respect her opinion on all sorts of matters. Based on her experience as a former high school teacher, she explained to Pauling and Freeman that the American education system had inadequately prepared high school students and instructors alike. As such, while a degree in education technically qualified individuals to teach all fields, Cooper had sometimes felt underqualified to provide instruction in subjects for which she lacked the necessary skillset. This anecdote helped Freeman to convince Pauling that writing a text for high school students wasn't a good use of his time as the content would too often fall into unprepared hands.

But most of Freeman & Co.'s projects found a happy medium between the sensational success of *General Chemistry* and the trapped potential of the Dickinson project, and overall Bill Freeman was pleased by his company's early progression. A milestone came at the end of 1949, when the publisher was able to put out its first catalog of books. Freeman was quick to point out that most houses were "in the red" for their first five years, but this was not the case for Freeman & Co. Rather, after only three years, the firm had published eight titles and favorably impressed numerous colleges with their small but successful operation.

Outside of their professional association with one another, Pauling and Freeman also forged a more personal bond. They visited one other as frequently as their busy schedules allowed, and their respective families also became close. In a letter to Pauling, Freeman remarked, "I gather that Lin [daughter Linda Freeman] fell in love with Crelly [son Crellin Pauling] even as Verne fell in love with Peter [son Peter Pauling]. I'll have to watch the women in my family around your boys." (Like Pauling, Freeman had a daughter named Linda. He and Verne also had a son.)

Pauling's immense faith in Freeman's abilities matched Freeman's deep respect for Pauling's life and work. The two often brought potential projects to each other, with Freeman suggesting at one point that Pauling star in a series of instructional science films. Who knows? Had he taken Freeman's suggestion to heart, Linus Pauling could have been the Bill Nye of the 1960s!

*

As Pauling's chemistry series matured, a third member of the team proved to be of crucial importance: Roger Hayward. Freeman had initially employed Hayward to create the illustrations for *General Chemistry* and, suitably impressed, wound up commissioning him for a number of other projects. The publisher quickly saw that Hayward was an incredibly talented illustrator and the two formed a close friendship that lasted through the years.

Freeman even recruited Hayward to design his new office when the company moved from Market Street. Freeman was color blind and admitted that he needed help when it came to outfitting a space. In the past, he had relied upon a "Vice-Presidents-in-Charge-of-Decoration" committee made up of several female colleagues, but ultimately decided that Hayward's instincts were a better fit. In particular, Freeman commissioned the illustrator to create a piece meant specifically to hang in his office. Hayward gladly obliged and, a few weeks later, presented Freeman with a large depiction of molecules that — as Freeman's staff informed him — were a lovely shade of apricot.

Professionally, Hayward was a key asset to the company, and in more ways than one. For example, when Freeman was courting John D. Strong, then at Johns Hopkins, for a book on optics, Hayward played a crucial role in convincing Strong to sign with the San Francisco firm rather than a larger publishing house on the East Coast. Once secured, Hayward and Strong worked well together, engaging in lively discussions as they collaborated on the manuscript.

As their relationship flourished, Freeman became a strong supporter of Hayward and his work. On one noteworthy occasion in the early 1950s, a textbook published by Houghton-Mifflin came under scrutiny because it contained a number of illustrations that closely resembled those that Hayward had created for use in *General Chemistry*. Hayward noticed immediately and brought the matter to Freeman's attention. Freeman replied with sympathy, suggesting that, "No one can copyright an idea. One can copyright the expression of an idea. This last applies, not only to the use of words...but to illustrations as well."

Believing that Houghton-Mifflin was indeed in violation, Freeman became aggressive. In the end, he negotiated a deal with the president of the company stipulating that Houghton-Mifflin pay a $0.04 royalty fee to Hayward for each copy sold.

Hayward's value was such that Freeman ultimately thought it wise to offer him a contract with the company, rather than enlisting him periodically as a freelancer. While beneficial to the company, this contract also offered the promise of a steady income and a new era of financial security for Roger and his wife Betty. The agreement seemed simple on paper. Hayward would continue to receive royalty payments for relevant books published prior to the contract, but for future work, he would receive an annual salary of between $4,500 and $7,500, depending on how many hours he logged. And though he would be on contract, Hayward would not be asked to be physically present at the office as long as he kept track of the time that he spent working on projects for the company.

These specifications were satisfactory for Hayward, particularly because of the royalties language, as continuation of these payments would insure some degree of income for Betty were Roger to pass away suddenly. But perhaps the most significant benefit of the agreement was the health insurance that the company offered to Hayward, who suffered from asthma. Importantly, the insurance agreement stated that Hayward and his family would receive health benefits during the run of the contract and also for two years after any event that led to his incapacitation or death. The only new requirement of Hayward was that he commit to completing illustrations for the exclusive use of Freeman & Co. and a partner publication, the *Scientific American* magazine.

After a period of negotiation, Hayward agreed to terms and, in 1958, began working as a company employee. His first task was to experiment with alternatives to color printing and texturing. While Freeman knew that color printing might help set his firm apart from its competitors, he worried that the costs were too high. Collaborating with Hayward, Freeman tinkered with screen tone and half-tone techniques, but neither proved especially successful.

Other professional complications emerged not long after. A year into his contract, Hayward began to feel that he was being treated more as a draftsman than an artist, and was too often compelled to engage in what he described as "uncongenial work." Freeman was open to Hayward's lament, but only to a point. "The boys around here," he wrote, "certainly have tried to see that each of us doesn't have to misuse his talents or have a disproportionate amount of the uncongenial. How in the world could we keep this spirit if we had an exception, one sharing in the perquisites but able to rule (by himself) that he would not do some share of the work because he decided it was uncongenial?"

But drudgery wasn't Hayward's only concern. Increasingly he felt that authors were overly critical of his illustrations and unwilling to give him the respect that he deserved. As Hayward grew increasingly unhappy, Freeman found himself spending more of his time mediating. This wasn't necessarily out of character; he often referred to himself as "Old Man Freeman," a persona of wisdom and benevolence that he felt he needed to embody as director of the company.

In Hayward's case however, assuming the persona began to wear thin. Circumstances reached a flash point when George Pimentel, who published *The Hydrogen Bond* with Freeman in 1960, put forth corrections to drawings that Hayward had provided. Upon learning of these suggestions, the illustrator took immediate offense and refused to complete the project. Freeman intervened to remind Hayward that authors greatly respected his skill as an artist but that his job as a company employee was to take direction from the authors or, at the very least, to negotiate with them. In response, Hayward wrote, "I see nothing in that [contract] which commits me to the philosophy that the author or any other person is always right. I certainly would never sign a contract which would require me to satisfy any person or persons who are not a party to that instrument."

Freeman's reply was terse. "We and our authors are the sole judges of what goes into a book," he wrote, "and you legally must abide by that principle or you do not fulfill your part of the bargain with this company." Suitably chastened, Hayward apologized to both Freeman and Pimentel and resumed illustrating.

With enough of these conflicts however, Freeman saw that a contract was no longer benefiting the authors, the company, or Hayward. Eventually he offered to terminate the agreement and allow Hayward to revert back to freelance status, receiving additional payments on a royalty basis. Hayward accepted and his relationship with the company, as well as his friendship with Freeman, quickly returned to a mutually respectful and co-productive state.

*

Pauling's landmark *General Chemistry* textbook hadn't yet cooled from the press when he, Freeman and Hayward began planning for a younger sibling, *College Chemistry*. Initially Freeman had expressed high hopes that Pauling might consider writing a chemistry text for liberal arts students and other non-science majors, but Pauling countered with another fundamental college text instead. He did so in part because he was troubled by several reviews that had dismissed *General Chemistry* as being too challenging for a freshman audience. Whether or not this was the case, Pauling felt that he could not adapt his existing book enough to adequately address the complaint. He also did not want to compromise his original intent, which was to give students who were passionate about chemistry a challenging and engaging textbook that would help them build a strong foundation of core principles within the discipline.

The gap between Freeman's hopes and Pauling's goals was ultimately bridged by *College Chemistry*, a text that specifically targeted the average student in its marketing. Intent on achieving this objective, Freeman insisted that Pauling's process incorporate a stable of editors who had absolutely no background in chemistry. Only in this way could the firm verify that this second book would be truly accessible to students lacking in specialized knowledge.

As Pauling circulated the first 15 and then second 15 chapters, Freeman became increasingly excited about their new project. Buoyed by these feelings and intent on using what he had learned from their experience with *General Chemistry*, Freeman laid plans for an early and aggressive advertising campaign. He and Pauling, however, had different ideas about the specifics of a successful sales strategy. In one instance, Freeman sent Pauling an ad that depicted nuclear fission with the following caption: "The above picture, of course, shows the process of nuclear fission. But it might also illustrate the way Linus Pauling's *College Chemistry* has hit American colleges. Has this chain reaction hit you?"

Pauling replied that the ad gave him the feeling of a text that had somehow disrupted American colleges, noting that "Whatever the disrupting effect is, it seems to be that it would be undesirable." Freeman quickly learned his lesson and steered away from science puns in future marketing attempts.

Despite their hectic schedules and a few setbacks, Pauling and Freeman managed to get *College Chemistry* published by 1950. The new book succeeded in reaching a wider audience than had *General Chemistry*, and the reviews that it received were largely positive. Pauling dedicated the book to his doctor, Thomas Addis, with the epigraph, "who in supplying science to medicine kept always uppermost his deep sympathy for mankind."

Shortly after sending the first edition off to press, Pauling was already working on revisions to *College Chemistry*. Though generally pleased with the work, he felt that there was room to further refine the writing and presentation of information. He also wanted to update the design. For instance, when he saw the color plate for a proposed illustration, Pauling was so impressed that he asked if they could print all of the second edition's illustrations in color. Though the newer technology excited him as well, Freeman denied the request, citing cost as a barrier. Such was the state of publishing at the time that Freeman felt they were already ahead of their competitors with the inclusion of a single color plate.

Despite the warm reception that *College Chemistry* received, Freeman warned that Pauling would have to make substantial revisions to further improve its accessibility while integrating new scientific developments that had come about between editions. Pauling heeded this warning and saved $300 of his royalties to recruit Fred Allen — then of Purdue University but formerly his professor at Oregon Agricultural College — to assist with the editing. At Freeman's request, Pauling also circulated his manuscript to non-science professionals, once again seeking to ensure that the text would connect with its target audience.

Another key change between the first and second editions was the royalty agreement governing Pauling's work for Freeman. As the company was still in its early years and in the midst of growing pains, Freeman proposed that Pauling receive a royalty of 10% for the first 10,000 copies (as opposed to their earlier contract, which stipulated a 15% royalty for the first 5,000 copies), 15% for the next 5,500 sold in a year, and 19% on all other copies sold in a year. Pauling agreed to these new terms for two reasons. First, it was his desire that the company's success continue. And second, the lower royalty rate meant that Freeman could sell the book at a lower list price. These compromises worked to increase distribution potential and, in the end, both author and publisher saw a satisfactory number of copies leaving the shelves.

*

As the 1950s marched along, W.H. Freeman & Company sought to actively build on past successes. In doing so, objective number one was to expand the number of textbooks that the company published. And for Bill Freeman, objective number two was to increase the firm's staff to match editorial and publishing demands.

One of the first employees that Freeman hired when he started his company was Janet MacRorie, the head of marketing. MacRorie and Freeman first collaborated on the advertising and marketing campaign supporting *General Chemistry* in 1947. In the early 1950s, Adam Kudlacik joined the firm as treasurer and secretary, thus beginning a lengthy tenure with the company. Harvey McCaleb also joined the team in 1953 to handle Midwest authors.

Perhaps the most significant addition to Freeman's staff came on board in 1949, when Stanley Schaefer joined Freeman and John Behnke as one of the firm's principal editors. The original intention for Schaefer was that he base himself in New York for purposes of recruiting and negotiating with East Coast authors, but Freeman was so impressed with Schaefer's work that he invited him to move to San Francisco not long after he was hired. Schaefer was pleased with the transfer and, in 1957, was promoted to executive vice president. (He remained with the company for several decades, eventually becoming president and chairman.) With Schaefer's promotion, another consequential hire was made when William Kaufman took Schaefer's old spot on the editorial staff.

Though business had been strong throughout the post-war period, by the late 1950s Freeman began to worry that his competitors were gaining traction. Looking to bolster his catalog, the publisher decided to redouble efforts to entice respectable authors to sign contracts with Freeman & Co. And while he continued to send outside manuscript proposals to Pauling for the chemistry series that he edited, Freeman also began to query Pauling for ideas on scientific areas that did not currently have a well-written or modern textbook in circulation. He then used this feedback to pinpoint his recruitment of authors to publish within those areas.

Freeman also updated the editorial plan for the book series that Pauling was heading. In particular, Freeman pushed the idea that non-science students could be harnessed to promote what he called "the revolution in scientific education." Likewise, because Pauling's *General Chemistry* text had been so successful, Freeman wanted to publish more non-traditional and experimental books in Pauling's line.

Pauling didn't disagree with Freeman's point of view, but he advised caution. Privately, Pauling was concerned that, in seeking to grow the company, Freeman might begin to adopt selection and recruitment policies that were employed by larger firms. Were he to do so, Pauling worried that Freeman might be tempted to sell the company if it grew beyond him. In Pauling's opinion, Freeman was fundamental to the company's success and this success could not continue — at least not in the same way — if Freeman allowed the company to pursue the same models as larger publishing firms.

Despite the hand wringing, W.H. Freeman & Co. flourished throughout the remainder of the 1950s. New heights were reached at the end of the decade, when Freeman announced plans to open a satellite operation in London. The establishment of this branch in 1960 opened new markets for the company across Europe, and did much to increase the firm's appeal among British authors.

Bill Freeman was beginning to receive recognition for his achievements as well. In 1960 he was awarded a Doctorate of Humane Letters from his alma mater, Hamilton College, and was also nominated by Pauling as California Industrialist of the Year. Though he didn't win this award, it meant a great deal to know that Pauling respected and admired him enough to nominate him for a prize that celebrated success through creativity and innovation.

Unsurprisingly, the company's reputation in its hometown was quite strong. On one occasion, the *San Francisco Chronicle* called Freeman & Co. "a great big sensational success story," providing ample evidence to back up the claim. At the time, textbook publishing was largely the domain of a handful of companies located on the East Coast. In fact, as the *Chronicle* article pointed out, there was only one reputable textbook publisher anywhere on the West Coast: W.H. Freeman & Co.

Amidst growth, change and strategic planning, Freeman's mission statement remained the same. The house would publish "only those [books] it thinks are based on a new and advanced viewpoint." But though the company's principles remained unchanged, its approach to publishing was becoming more experimental. Notably, as the 1960s moved forward, the firm entered into a joint venture with *Scientific American* to publish offprints of older articles, a partnership that would arise among shifting fortunes.

*

Difficulties at Freeman & Co. began to arise in the summer of 1959, when Bill Freeman separated from his wife, the former Verne Kopplin. When divorce papers were filed the following year, Freeman offered an even division of all their assets, with the notable exception of his stake in the company. Collectively, the couple owned 43% of the firm and Verne insisted that she retain her share. In her communications with Freeman, Verne pointed out that they had married in 1946, the same year that the company was founded, and that they had worked together to grow the company to its current stature. As such, she was entitled to a degree of control over its future direction.

From the outside looking in, Linus Pauling maintained a different point of view. In a letter to Freeman, Pauling expressed his feeling that Verne had done little to support the company beyond her contributions as a consultant. The possibility that she might retain a claim to the company was one that weighed heavily on Freeman. In a reply to Pauling, he revealed that "I find it quite impossible to carry on my work while sharing with her anything of my future." He also expressed concern that he might lose control of the company were Verne to retain her shares. Freeman knew that he would not be able to influence his ex-wife's decisions concerning the direction of the company. He also feared that she was planning to consolidate the stocks held by colleagues and friends to essentially buy the company out from under him.

Pauling did his best to provide a lift in his reply, writing, "I believe that W.H. Freeman & Company, as built up by you, has become the outstanding publisher of college textbooks of the highest quality in the United States... I was so greatly impressed by your ability that I felt that the advantage of having my book [*General Chemistry*] put out by your firm, because of your extraordinary ability and originality and convictions about the importance of publication of books of high quality, would outweigh the disadvantage of lack of an organization and reputation of long standing." He concluded that he wouldn't feel comfortable continuing his association with the company were Verne successful in reducing Freeman's control over it.

So strong was Pauling's conviction that he expressed a willingness to dramatically increase his skin in the game. Cognizant of the financial burden that the divorce and its aftermath had placed on Freeman, Pauling offered to buy Freeman's stock, which would provide Freeman with the capital to purchase Verne's shares should he wish. Freeman agreed to the proposition but only on condition that he be given the option to buy his stock back within three years. Pauling was not comfortable with this arrangement and the two failed to arrive at a solution that would satisfy them both.

In the end Verne retained her shares, and once the divorce was settled in the fall of 1960, Freeman continued to spiral. In order to keep Verne from gaining control of the company, he was obliged to purchase at least 200 of her shares at $55 each while also paying the mortgage on the house that they had owned together. In a letter to Roger Hayward, Freeman bemoaned his state of affairs: "Old Man Freeman feels like the tempest in a terribly small teapot; no one ever gives any thought to the tempest's feelings or understands how constrained he feels."

Needing an escape, Freeman took the summer off to travel around Europe. He made it as far as Greece and self-published a book describing his experiences, titled *Ola Kala: The Greek Word for It*. Meanwhile, tensions mounted at W.H. Freeman & Co. as its eponymous leader seemed increasingly unstable. A growing sentiment among many stockholders was that Freeman would do anything to keep control. As this idea gained traction, executive vice president Stanley Schaefer became nervous about the future of the company and sent out a request to many of the stockholders that they become proxies, thereby granting them the authority to make decisions about the company.

Finally, in January 1962, Bill Freeman agreed to sell his stock, though he was resistant to sell within the company because of his objections to the firm's recent association with *Scientific American*. (It is likely that the arrangement with *Scientific American* was entered into to provide a measure of protection for the company amidst the financial damage caused by Freeman's divorce.) In his correspondence with Pauling — one of the few people at the company that he still trusted — Freeman railed against the decision and expressed sharp criticisms of Stanley Schaefer and Bill Kaufman as well as other long-timers like Harvey McCaleb and Adam Kudlacik. Pauling balked at these denunciations, pointing out that Freeman had hand-picked these men and needed to trust in their judgment, as Pauling did.

After a different and particularly troubling discussion with Freeman, who sometimes met the Paulings for dinner, Linus reflected on the current state of the company, noting that "Bill and Verne damaged it, neglected it, and [devoted] their energy to fighting each other." Though he was sympathetic to Freeman's situation and deeply concerned about his friend, Pauling believed that there was no justification for the pain that Freeman was causing the company.

Stanley Schaefer also wanted to help and, like Pauling, offered to buy Freeman's stock. When Freeman declined, Schaefer suggested that *Scientific American* could purchase the shares instead. Freeman felt that this was not a realistic solution either. He did, however, agree to not sell his stock to a competing company. When Freeman subsequently took a job at Addison-Wesley's western office, signing a contract that would allow the company to purchase his Freeman & Co. shares, he effectively broke this promise.

When Pauling asked Freeman why he had done this, Freeman confessed that he was too dissatisfied with the present management at Freeman & Co. to consider associating with it anymore. *Scientific American* stepped in at this point and made an offer for Freeman's stock that Addison-Wesley could not match. Ava Helen Pauling, who remained a confidant for Freeman, advised him to sell his stock to the magazine publisher. Doing so, she reasoned, could secure a stable future for his children while also providing an avenue for Freeman to leave his old company gracefully.

Freeman reluctantly agreed and Gerard Piel, the president of *Scientific American*, put the money from this transaction into a trust fund. Trustees "in whose rationality and integrity" the company had confidence would vote at a later date on the matter of what to do with the proceeds. Meanwhile, once her ex-husband had become associated with Freeman & Co. in name only, Verne also lost interest in company matters and refocused her energies on her legal career. She eventually remarried and went on to challenge discriminatory policies at law firms in Connecticut, where she practiced for several years.

Sometime later, having relocated and newly married to his former secretary, Margaret Cooper, Freeman reached out to Ava Helen to explain his behavior. In his letter he confided, "For old times' sake, I will say to you that I had no alternatives — financial ones possibly, but professional or personal ones, absolutely none... As I've said to Linus, until the future speaks, I trust that we can each of us respect the other's right to act in accordance with his convictions."

*

Despite the disappointing end to his involvement with the company that he had founded, Bill Freeman worked in the publishing industry for the rest of his life. Post Freeman & Co., he stayed with his new employer, Addison-Wesley, long enough to regain a sense of confidence. With his second wife, he then set about establishing a new independent press: Freeman, Cooper & Company, a name that once again incorporated the Freeman brand, but now also included Margaret's maiden name.

Rather than limiting the scope of their new firm to a specific discipline, Freeman, Cooper & Co. published books on a wide range of subjects. And while he was still primarily interested in textbooks, Freeman shied away from entering into direct competition with his previous company. As a result, the new venture published books for academic use that were not, strictly speaking, textbooks. And while science remained a key area for the publisher, other subjects including psychology and philosophy also began to populate the catalog.

Though Freeman relished the fresh start, he still recognized the value of retaining past connections. Key among these was his close friend Roger Hayward, whom Freeman approached with a few project ideas in 1971. The first of these was a book that he hoped Hayward would write on "the simpler and fundamental geometry of nature," intended for use by both introductory and advanced students. He also proposed that Hayward illustrate a different title on crystallography for chemists, and a third book focusing on the chemical elements.

Hayward expressed interest in these projects, as long as Freeman could pay him royalties. Freeman agreed, but warned that the income might be small because the audience for each project was likely to be rather specialized. For Hayward, this was a risk worth taking, given that his earnings from other projects were robust enough to absorb a potentially low payout from these new ventures. Having arrived at this understanding, Freeman's only additional request of Hayward was that he not complete illustrations for any rival publishers. This was also an easy ask to fulfill as Freeman, Cooper & Co. were engaged in direct competition with only a few other firms.

Their agreement in place, Hayward set to work on new illustrations and an early draft for his geometry of nature book, contacting Freeman regularly to keep him apprised of his progress. By this point in his career, Freeman no longer held his authors to strict deadlines, so long as they did a good job of staying in touch. In their exchange of letters, Freeman provided gentle guidance while Hayward developed his text. When Hayward broached the idea of including anecdotes from his personal life in the book, Freeman expressed reluctance. And while he ended up telling Hayward to proceed, he advised caution: too much autobiography could harm an author's academic authority, he felt, though the right amount of personal narrative could work to forge a deeper connection with students.

Once Freeman had piqued Hayward's interest with these smaller projects, he unveiled the idea that he was most excited about, an organic chemistry manuscript penned by Ruth Walker, a chemist at Hunter College. Hayward enthusiastically agreed to provide illustrations for the text, but soon became enmeshed in a familiar set of struggles: when Walker raised concerns about Hayward's initial drafts, the illustrator refused to make changes.

Most of Walker's concerns were over small or superficial details in the illustrations but, oddly enough, a particularly contentious debate eventually led to a significant advancement. As part of his portfolio for the project, Hayward had created a paper model of a tetrahedron that was designed for students to tear out and construct into their own molecule. While on board with the idea, Walker claimed that the instructions that Hayward had written were inaccurate and that the overall design was ineffective.

Unable to resolve the debate themselves, Walker and Hayward brought the matter to Freeman. The publisher was intrigued by Hayward's unique design, but agreed with Walker that it would be difficult for students to follow the instructions that Hayward had provided. As a means of clarification, Freeman suggested a minor modification — the addition of dotted lines to indicate the direction in which students should bend the model. He also promised Hayward that he would collect more authoritative opinions from accomplished chemists and reconvene with him before the publication of the text.

Several of the chemists whom Freeman contacted agreed that Hayward's model was unique and had potential as a teaching tool. When Freeman relayed this feedback to Hayward, the illustrator immediately took steps to patent his design. Freeman assured him that the copyright protecting the material in the book would be sufficient for this purpose as well.

Once it was established that Hayward and, to some extent, Freeman had created something new, Freeman's associates in the scientific world went about naming the structure. They eventually settled on the Cooper Structure, an obvious source of frustration for the illustrator. Peeved, he crafted a short memo to Freeman that was written in the large, bold typeface that he had adopted as a result of worsening eyesight. "For goodness' sake," it read, "what's wrong with the Hayward Structure?"

Freeman replied that the Hayward Structure was actually the first name that had been proposed, but that the group of scientists couldn't arrive at a consensus. Some other names they tried included the Freeman Structure, a hopper crystal, a starved tetrahedron (because of the model's concave sides), an inverted dodecahedron, an instellated polyhedron, a Texas Tetrahedron, and a Cooper Crystal. The Cooper Structure was the name that everyone ultimately agreed upon. Hayward belatedly suggested the HFC Form — for Hayward, Freeman, Cooper — but his submission was largely ignored.

Changing tactics, Hayward once again began investigating a patent, arguing that copyright protections simply prevented anyone from publishing the design. Freeman commiserated with Hayward's feelings, but was firm in his resolution that a patent was not necessary. In fact, because Freeman's modification is what made the model effective, Ruth Walker gave him credit for its discovery. In a *Journal of Chemical Education* article, she wrote,

> "A unique model for illustrating the tetrahedral geometry of sp³ bonding is obtained when the pattern in the figure is cut out and assembled...the resulting structure is a tetrahedron with four recessed faces and a central hole, and has been named the Cooper Structure. Each face is recessed in such a way as to produce a model that clearly shows the relative position of four bands extending from the center of a tetrahedron, one towards each apex. This model was designed by William H. Freeman for inclusion in 'Organic Chemistry: How to Solve It (I. Molecular Geometry)' by Ruth A. Walker, after Mr. Freeman observed models made by Roger Hayward, the illustrator of the organic workbook published by Freeman, Cooper & Co. in 1972."

Hayward was somewhat placated by the wording of the article, which let him claim a share of the credit for the design. He proceeded to recommence work on his geometry of nature book, but never finished it as his health problems increased in severity.

Meanwhile, Bill Freeman was also experiencing his share of setbacks. Just before the Cooper Structure conundrum arose he was hospitalized for exhaustion, which slowed production considerably and led to a period of prolonged discouragement. In a letter to Hayward, he made reference to "hurdles, disappointments, problems and shenanigans that I dare not put into print." And for all the fuss that it caused, Ruth Walker's book, *Organic Chemistry — How to Solve It*, sold only 11,000 copies.

Over the course of its history, Freeman, Cooper & Company experienced moderate success, but never achieved the status of its predecessor. Many of the authors who had found their niche at W.H. Freeman & Co. remained loyal to the original company even after its namesake had moved on; indeed, with the notable exception of Roger Hayward, Bill Freeman built his new company largely from scratch. He insisted, though, that modest successes did not diminish his passion for the independent press. After he passed away in 1992, Margaret took over the firm and ran it efficiently for a few additional years before letting it go to become another piece of publishing history.

*

In the years immediately following Bill Freeman's departure from the company, Stanley Schaefer ran W.H. Freeman & Co. quite smoothly. In 1969 Bill Kaufman took over as president with Schaefer staying on as chairman, and Kaufman also did well. Notably, he played a key role in the release of revised editions of Pauling's *General Chemistry* and *College Chemistry*, and by the end of his first year in charge, Schaefer was able to report that the company had grown. As the onset of the 1970s loomed, Freeman & Co. had published 14 new books and added 72 titles to its *Scientific American* offprint series. The outlook for the next fiscal year seemed bright.

The connection with *Scientific American* was especially important, as the company had formally merged with the publication in 1964. Of this change Schaefer remarked, "Now united are the forces of two successful, non-competitive publishers who have outstanding reputations for high standards and excellence in scientific publishing. Each is making distinctive contributions to the new alliance. I mention, for example, the significant new source of authors for Freeman books that is now available to us." Roger Hayward, who had spent years working for both Freeman and *Scientific American*, expressed surprise at this news, but congratulated both parties and noted that the transition seemed to him a "happy circumstance."

That same year, Pauling and Hayward began collaborating on *The Architecture of Molecules*, originally titled *Molecular Architecture* but renamed by Pauling just prior to its release. A stunning and unique collision of science and art, the book found immediate success and continued to do well for years afterward. Both collaborators received 15% royalties for the first 10,000 copies sold in cloth, 18% for every cloth-bound copy sold beyond the initial 10,000, and 10% for the paperback editions.

Despite his earlier claim that he would not feel confident in the company without Freeman directing it, Pauling continued to maintain a positive and productive relationship with his long-time publisher. In one particular instance, Pauling played an instrumental role in smoothing tensions caused by an unflattering review of a Freeman text. The review in question was authored by microbiologist C.B. Van Niel, whose highly critical assessment of Wayne W. Umbreit's *Modern Microbiology* appeared in the widely read journal, *Science*.

Prior to the review appearing in print, Van Niel had sent a letter to Bill Freeman warning him that it would not be favorable, but Freeman had left the company by this point. His replacement, Stan Schaefer, didn't see the piece until it had been published and, upon reading it, responded personally to Van Niel, writing that the criticisms had hit sales hard. Schaefer further speculated that Van Niel harbored a personal grudge against Umbreit and that this was the real source of the animus permeating the review.

It was at this point that Pauling came to Schaefer's aid. He informed Van Niel that he personally had not found the book to be nearly as flawed as the review claimed and accused his correspondent of "malicious mischief," stating that most of the errors that he attacked were simple and relatively common across publications. Without waiting for Van Niel's response, Pauling then wrote to Phil Abelson, the editor of *Science*, asking him to retract the review because it was disrespectful, incorrect, took sentences out of context, and was overly aggressive in tone. Aligning with Pauling's defense, many other professionals in biology and bacteriology spoke out against the review, criticizing its focus on minor errors. More importantly, many within this group also chose to adopt the text, despite its flaws.

To stave off future conflicts of this sort, Schaefer requested that, as a courtesy, drafts of reviews be sent to Freeman & Co. before publication, so that the company could prepare if the analysis was unfavorable. Pauling also asked Van Niel for his own annotated copy of the Umbreit text so that Umbreit could use it in his revision process.

When Schaefer promoted Bill Kaufman to president in 1969, he assumed the position of chairman, a post that had previously been occupied by Freeman himself. Kaufman opted for early retirement in 1971, telling Pauling that the timing felt opportune because the "fame" of the company was at an all-time high. He was also confident in the competence of the staff and its collective motivation to ensure the continued success of the company.

Pauling was also feeling bullish about the company's prospects, so much so that he finally brought up an issue that had been troubling him for some time. Contractual modifications that Bill Freeman had instituted for the second edition of *College Chemistry* — modifications that lowered Pauling's royalty rate — were presented as temporary changes needed to help grow the young company financially. When it was suggested, Pauling saw no problem with the change, so long as it was temporary. But as far as he could see, the lower royalty rate had been applied to the third edition of *College Chemistry* as well, and Pauling came to feel that he was being taken advantage of. In a letter to Schaefer he expressed his feeling that the agreement, as it was being continued, "might be said to have been obtained by fraudulent methods, involving statements to me that I think were untrue or at least misleading about the financial situation of the Company."

Schaefer checked the royalty statements and concluded that Pauling was correct in his assessment. After apologizing and thanking Pauling for bringing the matter to his attention, he then set about calculating the difference between the royalties Pauling had received and the royalties that should have been dispensed. Once done, Schaefer assured Pauling that the company would pay him $5,000 owed for the second edition of *College Chemistry* and $7,400 for the third. Pauling thanked Shaefer for his straight dealing and then requested that the company pay him interest at the rate of 7% on these remittances, because they were late.

*

In the 1970s and 1980s, well after Bill Freeman's departure, Pauling published a number of other books through Freeman & Co., including two big sellers, *Vitamin C and the Common Cold*, and *How to Live Longer and Feel Better*, as well as *Orthomolecular Psychiatry*, which was more of a niche volume. But as the staff at Freeman & Co. evolved, Pauling began to experience trouble communicating.

Over time, Pauling also felt his editorial contributions were being restricted. Notably, when *Basic Physical Chemistry for the Life Sciences* was published as part of his series, Pauling had no hand in editing it. Once the book had gone to the printers, Pauling sent a letter to Stanley Schaefer stating his belief that the text should not have been released in this manner. No title, he felt, should be published in his series if he was not really and truly the editor.

Schaefer replied that he had no intention of impinging upon Pauling's authority over the series, but did express his feeling that Pauling's fame worked against the company at times. In addition, Pauling's manuscript comments were often very blunt, and while it is unclear how many authors were given the opportunity to read these assessments directly, those who did were often upset by comments that they felt were unfair. One author in particular, an R. Nelson, wrote several pages to Schaefer defending his stylistic choices after the company had rejected his manuscript. Pauling, for one, had criticized his informal tone, but Nelson felt that the approach made the book more appealing to younger generations of scientists. Nelson then attacked the company's decision to retain Pauling as an editor for the chemistry series, writing, "The problem really arises because the chemistry editor is the author of the first text and is a man of strong convictions (as well as great prominence). I believe that this situation puts a potential author (one with no prominence) in an untenable position."

Schaefer was moved by this comment in particular and asked Pauling to reconsider the manuscript with the understanding that some minor errors would be corrected. In so doing, Schaefer also sided with Nelson's point of view in suggesting that Pauling's take was based largely on differences in style. In this instance, Pauling was flexible and reconsidered.

In 1970, to help with the problem that Nelson had raised, Schaefer hired Pauling's friend, colleague and former student, Harden McConnell, to serve as a co-editor for the chemistry series. McConnell was someone whom Pauling respected and who also tended to be rather more gentle in his critique, and the arrangement worked out well. The collaboration likewise helped to spread the editorial responsibilities such that Pauling could dedicate more time to other projects.

Circumstances had changed substantially by 1979 when Richard Warrington, the latest president of the company, suggested that Pauling terminate his editing contract. Though he stressed that Pauling's "association with the company as an author and adviser in the early years was very important to the success that followed," Warrington also pointed out that Pauling was no longer teaching. As such, Warrington worried that Pauling's interests and priorities had changed significantly. He also felt that the Freeman company hadn't been as effective at bringing in successful chemistry texts in recent years. Pauling felt similarly, noting that the flow of manuscripts from the company had slowed considerably.

As the company continued to experience leadership changes throughout the 1980s and 1990s, Freeman's relationship with Pauling as an author also began to deteriorate. Notably, around the same time that Warrington had asked him to terminate his editing contract, Pauling discovered that the company had allowed *General Chemistry* to go out of print. Another shift came about a year later when Neil Patterson assumed the role of president and moved the company to New York to be closer to *Scientific American*, with which Freeman & Co. now shared a CEO. Pauling had long enjoyed having a publisher based on the West Coast and was disappointed by the move.

In 1991 a correspondent named Jonathan Paul Von Neumann wrote a letter to Pauling expressing his own disappointment that

Freeman & Co. hadn't shown any interest when he approached them with an offer to translate *How to Live Longer and Feel Better* into other languages. Pauling wrote back, sharing Von Neumann's concern and confiding his belief that publishing companies often mishandled their authors.

The next year, Pauling's relationship with Freeman & Co. all but came to a close when the publisher rejected two of his proposed manuscripts. One was a freshman-level text that he planned to write with his youngest son, Crellin. But perhaps more disappointing was the firm's lack of interest in Pauling's second suggestion, a memoir that he was to title *The Nature of Life — Including My Life*.

W.H. Freeman & Company was eventually reabsorbed into its parent company. Now an imprint under Macmillan Learning, the group continues to publish successful science textbooks and provide other educational resources.

Christmas

Chapter 4

By Linus Pauling Jr.

Linus Pauling Jr. celebrating his first Christmas, Pasadena, California, 1925.

[As an undergraduate] at Pomona...I was serving dinners to some of my classmates, these big trays loaded with plates. They were prepared in the kitchen and carried out and distributed around. So I was pretty good at that. Anyway, in the dining hall — which was the famous Frary Dining Hall at Pomona which has a huge Orozco painting of Prometheus which is famous in art circles — the dining room brought in a beautiful Christmas tree about 20 feet high with delicate pine cones on the branches. A few days before Christmas, the college emptied out of course, as everybody went home. Since I was a member of the dining room staff, I said,

"What's going to happen to the tree?" "We're gonna throw it out into the dumpster." So I said, "Can I have it?" "Sure."

So my '32 Ford Roadster had a windshield that would fold down flat against the hood. I drove up to the dining room, got the tree, laid it out across the car — and you know it was longer than the car was — and beside me in the passenger side with the windshield down, drove back to Pasadena with this tree on my car and set it up in the living room in the Fairpoint Street house. The living room there had ceilings that were about 12 feet high which was pretty good, but not quite good enough for this tree. I had to chop off some of the top and some of the bottom to get it to fit, but it was probably the most magnificent tree they ever had in there.

The traditional Christmas was that, when I was a kid, the household would go on absolutely normally through Christmas Eve with no sign of any particular holiday spirit. Then somehow, on Christmas morning, I'd wake up and here was the tree fully decorated with the lights and presents and so on. I wondered how this happened. Finally, when I was six or seven, I got old enough to realize that somebody had done all this, and then I became party to the adventure.

It turned out that on Christmas Eve, my father would go out late in the evening. By then, all of the itinerant Christmas tree salesmen had left, leaving their leftover trees in the lot to be thrown out. So we'd roam around Pasadena looking at all of these trees in these now-deserted lots and find one we liked, take it, put it on the car, and drive home. Then there'd be frantic activity through the night to put on the lights and the decorations, the ornaments, and bring out the presents for the younger siblings to wake up to and be amazed at. So the same thing, I presume, happened to them at the right level. All that came to a halt finally, probably with Crellin being old enough. That was really exciting and one of the fun parts of Christmas. I never followed that with my own children. Christmas got going early and they were exposed to it for a longer time.

The [Pasadena] tree stayed up until after New Year's. It was too nice, too beautiful, to think about taking down right away.

The Soviet Resonance Controversy

Chapter 5

By Miriam Lipton

"Glory and Pride of Russian Science," *Pravda*, November 22, 1961. Note Pauling's annotation at far right indicating his presence on the dais.

Linus Pauling's theory of resonance helped to unify the classical roots of organic chemistry with the new field of quantum physics. In so doing, the theory provided a hugely important framework for understanding observed atomic behaviors that did not correlate with the era's prevailing mathematical models of the atom. The theory would also help to usher in an onslaught of new approaches to organic chemistry, lifting, in Pauling's words, "the veil of mystery which had shrouded the [chemical] bond during the decades since its existence was first assumed." It was, in short, one of the most adaptive and applicable postulates ever put forth by Pauling.

But the theory of resonance was not immune to controversy, receiving early resistance from within the U.S. scientific community and, in a very different way, from abroad in the Soviet Union. The disputes surrounding the theory were relatively short-lived though, and Pauling's ideas on resonance continue to inform today's understanding of molecular architecture.

Pauling's ideas on resonance were grounded in the work of several scientists but most crucially August Kekulé and Werner Heisenberg, both of whom were also interested in the structure of molecules. Kekulé (1829–1896), a German chemist, notably devised a proposed structure for benzene, an aromatic hydrocarbon of interest to many. Kekulé's model put forth a structure consisting of six carbon atoms forming a ring, with hydrogen atoms attached externally to each carbon. Though intriguing, this basic structure did not explain where, on the interior carbon ring, double bonds were located. Partly because of this, Pauling would later lament that "the Kekulé structure for benzene is unsatisfactory."

Shortly after Kekulé published his basic benzene structure, multiple isomers — or alternative structures — of the same compound were predicted and even isolated by Kekulé, but even these breakthroughs were not enough to confirm the "correct" model of benzene. Recognizing as much, Kekulé later posited that, in actuality, benzene "oscillates" between the various isomers, and that all isomers may in fact be regarded as "correct." This notion of oscillation between isomers was hugely important, but despite its utility, Kekulé never succeeded in accurately predicting the "true" structure of benzene. Instead, the solution to the puzzle would lie in waiting for nearly 60 more years and would rely heavily upon Pauling's resonance breakthrough.

In the meantime, Kekulé's oscillation theory served well in disrupting traditionally held beliefs about the number of valence electrons that must be present in aromatic compounds. This, in turn, helped to usher in new theories about the chemical structure of aromatic compounds more generally. By the 1920s, a community of American, British and German chemists had developed a set of theories related to aromatic compounds that built on Kekulé's ideas. The group's basic hypothesis was that, instead of molecules oscillating between various isomers, perhaps all isomers actually existed simultaneously? This idea of simultaneous existence piqued Pauling's interest because it seemed related to work that he was doing with quantum mechanics — specifically, ideas related to quantum resonance that had been introduced by Werner Heisenberg in 1926.

Heisenberg (1901–1976), a contemporary of Pauling's, was working to understand the wave mechanics of subatomic particles. As part of this project he theorized that, on the subatomic level, molecules exist in discrete quantum states and that the wave function of a given molecule can be described as the sum of these various states. Heisenberg coined the term "resonance" to refer to this process — e.g., the summation of various quantum states to comprise a molecule's wave function. Pauling was intimately familiar with Heisenberg's theory of quantum resonance as well as the hypotheses proposed by the British, American, and German contingent. Thus equipped, he began to construct a theory of his own that would prove crucial to building a more robust understanding of molecular architecture and chemical bonding.

Pauling circulated his resonance theory in a series of papers that were published from 1931 to 1933. In them, he considered the ideas stated above, before emphasizing that "the actual normal state of such a molecule does not correspond to any one of the alternative reasonable structures, but rather to a combination of them, their individual contributions being determined by their nature and stability." In other words, the individual isomers of a given molecule should not be viewed as existing in a state of rapid switching from one to another. Instead, a hybrid of every isomer is, in fact, the "true" form of the molecule.

The distinction that Pauling drew between rapidly switching isomers — a state known as tautomerism — and isomer hybrids was conceptually difficult for many of his peers to grasp, but Pauling was able to cite experimental evidence to support his theory. Namely, Pauling had found that resonating molecules existed at a much lower energy state than tautomerism would predict. Pauling believed that these lower energy states resulted in more stable molecules, an effect that lent support to the viability of resonance — as opposed to tautomerism — as an operating theory.

The experimental data continued to be important to Pauling as he pushed his theory forward. Some had argued that there was no real difference between resonance and tautomerism, because the classical understanding of tautomerization portrayed isomers as switching so rapidly as to be in a virtual hybrid state of their own accord. But the data showed that Pauling was describing something different and that, to use Pauling's words, "it is easy to distinguish between the two."

In a 1946 speech delivered to a private industry group, Pauling restated the basics of his theory using language that is useful for summarizing here. For a hypothetical molecule known to have two isomers, "neither the first structure nor the second structure represents the system. Instead, the molecule is 'a combination' of the two structures." Further, when scientists "can write two structures, neither one actually represents the state of the molecule, but both of them together represent the state of the molecule. The molecule," he concluded, "is more stable actually than it would be if it had any of the structures that you can assign to it."

Pauling was eventually able to persuade most of his colleagues to align with his thinking on the theory of resonance, in part by using the theory to solve the elusive structure of benzene. One of the reasons why chemists knew that Kekulé's model was incorrect was because the observed energy level of the benzene molecule was much lower than the number that Kekulé would have predicted. Something else, then, was causing the energy of benzene to be lower (and thus more stable). Pauling's theory posited that resonating hybrids exhibit lower energies, and ultimately he was able to use his ideas to build a structure of the molecule that fit with the energy data. Once the model was accepted, the benzene breakthrough did much to secure resonance theory as a valuable tool for understanding molecular structure.

*

As it evolved, resonance theory came to inform the two schools of thought then prevailing for chemists interested in quantum mechanics and the chemical bond. These two modes of thinking were the Valence Bond (VB) theory and the Molecular Orbital (MO) theory. And while both theories used the concept of resonance to explain molecular structure, neither was able to predict all structures all of the time.

One might assume that this variability would have troubled Pauling, but if so, it did not cause him to falter in his conviction that resonance was the best tool available for accurately describing molecular states of matter. Perhaps because of this, Pauling ultimately preferred the VB theory, for which hybridization of orbitals is the mechanism by which molecular shape is explained. That said, Pauling was also aware of the utility of MO theory, and in at least one instance — a study of cyameluric acid — he was compelled to use MO theory to explain the structure, having concluded that his application of VB theory was inadequate.

But incidents like this did not weigh heavily on Pauling, who remained supremely confident in the soundness of his ideas. Notably, in *The Nature of the Chemical Bond*, Pauling explained away the incongruities between VB and MO theory and upheld resonance by stressing that, "the convenience and value of the concept of resonance in discussing the problems of chemistry are so great as to make the disadvantage of the element of arbitrariness of little significance."

While he took pains to press the utility of resonance theory, Pauling sometimes found it more difficult to explain its vagaries. On a conceptual level, one of the theory's biggest drawbacks was its tenet that the "correct" structure of a molecule is a hybrid of all possible isomers of the molecule. This effectively meant that, in a resonating structure, all possible isomers exist simultaneously, an idea that proved confounding for many. Where prior to resonance the conventional wisdom was based on Kekulé's description of rapidly changing isomers, Pauling was sure that, in order for resonance theory to work, the idea of isomers shifting from one to another had to be abandoned.

In making the case for resonating structures, Pauling wrote that "a substance showing resonance between two or more valence-bond structures does not contain molecules with the configurations and properties usually associated with these structures." Of the various Kekulé structures, Pauling then suggested that "taken together, [they] provide a rough description of the wave function of the molecule in its normal state." The implication, of course, was that resonance theory offered a more precise set of tools for understanding molecular architecture, once one was ready to clear a few conceptual hurdles.

In time the initial fog of confusion began to lift, and the practical brilliance of Pauling's theory led to its widespread acceptance. Indeed, while some saw resonance as heralding a promising new direction for the application of chemistry, others found the theory appealing simply because it seemed to solve so many problems. Early praise came in 1933, when the esteemed British chemist Christopher K. Ingold declared the theory as having effectively resolved lingering questions about the stability of aromatic compounds. Ingold's colleague Nevil V. Sidgwick also became a big supporter of resonance theory. Sidgwick was a well-respected scientist, and his endorsement of resonance helped to cement its status as a leading model for understanding chemical structures. A few years later, in his 1944 book, *The Theory of Resonance and its Applications to Organic Chemistry*, University of Chicago professor George Wheland proclaimed resonance to be "the most important addition to chemical structural theory that has been made since the concept of the shared-electron bond was introduced by G.N. Lewis."

Pauling too believed in the utility of resonance throughout his life. He argued in particular that the theory was especially well-suited to "aromatic molecules, molecules containing conjugated systems of double bonds, hydrocarbon free radicals, and other molecules to which no satisfactory single structure in terms of single bonds, double bonds, and triple bonds can be assigned." He also believed that "resonance provides an explanation of the properties of many inorganic molecules. For example, the carbon monoxide molecule." Pauling later reflected that resonance theory "permitted the discovery" of the alpha helix and associated models comprising "the most important secondary structures of polypeptide chains in proteins."

*

But as it turned out, not all found comfort and solace in resonance. In particular, many Soviets viewed Pauling's theory as an affront to their ideology, regardless of its scientific "correctness." For these critics, an abstract hybrid model of the molecule — one lacking a foundation in a quantifiable reality — was too troublesome to ignore.

Pauling's resonance papers were published in the early 1930s and expounded upon at the end of the decade in *The Nature of the Chemical Bond*. And while Soviet scientists were surely aware of resonance theory throughout this time period, it appears that ideological interest in the work did not fully manifest until after World War II. This was likely because, during the war, the Soviet government diverted scientific initiatives away from theoretical investigations in favor of practical applications that would benefit the war effort. Shortly after World War II ended however, Soviet scientists Ia. K. Syrkin and M.E. Diatkina translated *The Nature of the Chemical Bond* into Russian. The tandem then wrote their own book outlining ideas on structural chemistry, a book that was based on resonance theory.

Though they had grounded their text in well-established chemical theory, Syrkin and Diatkina's work was almost immediately criticized within the USSR. In 1949, Soviet scientists V.M. Tatevskii and M.I. Shakhparanov published an article in the journal *Voprosyi Filosofii* (*Questions in Philosophy*) titled, "About a Machistic Theory in Chemistry and its Propagandists." In it, the authors contended that Syrkin and

Diatkina were obliged to carefully critique Pauling's resonance theory, rather than bestow upon it "laudatory" praise. The paper also pointed out that Syrkin and Diatkina did not include "even one mention of the works of Russian or Soviet scientists. [I]t was expected that the translator and editor," they felt, "fighting for the honor of the Soviet science, would have seen to it to fill in this gap."

"Still worse," the authors continued, "the annotations which Syrkin gives at the end of the book are represented primarily as indication of the works of American and English chemists." This given, one could only conclude that Syrkin and Diatkina were merely "propagandists for the known-to-be erroneous and vicious theory of the American chemist, Pauling."

Tatevskii and Shakhparanov's views were not anomalous. In fact, the Soviet Academy of Sciences itself viewed Syrkin and Diatkina as being in "serious error" for "championing the theory of resonance or failing to expose its supposed fallacy." The rhetoric only escalated from there, and in 1951 Syrkin and Diatkina were obliged to recant their belief in resonance theory. In doing so, Syrkin was famously quoted as saying that he had "overestimated second-rate works of foreign scientists."

On the surface, the criticism of Syrkin and Diatkina seems to have emerged primarily from their support of a non-Soviet scientist. And while this was certainly a factor, the attack was also a product of a larger ideological push within Soviet science that traced its origins to the 1930s and an individual named Trofim Lysenko. An agronomist, Lysenko came to symbolize the ideal Soviet man, a notion that received the full support of Premier Joseph Stalin. Lysenko had been born into relative poverty but eventually gained entry into the Kiev Agricultural Institute, where he studied wheat production. It was there that he began to develop the ideas for which he would become well-known.

Lysenko believed that if he could freeze winter wheat, he would be able to force early germination of the crop, a process known as vernalization. He also believed that this early germination would in turn become a heritable characteristic. A breakthrough of this sort was highly desirable in the Soviet Union, in part because the vast nation had experienced multiple food shortages due to unseasonably cold winters and poorly structured collectivist agricultural practices. In Lysenko, Stalin saw the promise of more easily feeding the Soviet population, and the scientist quickly grew in stature as a "man of the people."

But Lysenko's thinking was attractive on ideological grounds as well, chiefly because his ideas on vernalization took a Lamarckian approach to evolution. Simply put, Lysenko believed that evolution was something that could be controlled by man; that the people had the means and ability to enact and propel genetic change. This idea ran counter to those prevailing in other parts of the world and specifically the United States, where Darwinian evolution had been embraced. In the Darwinian model, evolutionary change occurs by random chance, and individuals do not possess the ability to create lasting heritable impact through their own actions. For the Soviets, this concept was seen as bourgeois, a word that was widely held in contempt. Lysenko's contrasting view of evolution was one that put the state in control, and it was absolute.

Given this context, it becomes easier to see how resonance could be viewed as being counter to Soviet ideology. And within the USSR, an alternative theory was accordingly developed by a Soviet chemist named Gennadi Chelintsev, who aspired to become the chemistry and biology equivalent of Lysenko. Chelintsev applied Soviet ideology to chemistry in his book, *Essays on the Theory of Organic Chemistry*, using the text to argue that every molecular compound had only one discrete structure. His ideas were appealing to many Soviet scientists because, as with Lysenko, they were primarily based in the mechanistic worldview that was so deeply enmeshed within Soviet society.

*

Throughout the 1940s, the culture of Soviet science became ever more dominated by Lysenkoist thinking, with little room permitted for other points of view. Any work that did not fall strictly in line with Lysenkoist ideology was condemned, and those scientists who were brave enough to uphold contrary beliefs were routinely imprisoned or even killed. This period, which coincided with the end of Joseph Stalin's reign as premier, was clearly not an environment that encouraged innovation; rather, conformity ruled the day.

By extension, the most lauded Soviet scientists of the era were those who remained ideologically attuned to accepted beliefs, and it was in this environment that pitched opposition to Pauling's resonance theory emerged. The controversy, then, was not so much one based on scientific opposition, but was instead steeped in perceived ideological differences.

Tatevskii and Shakhparanov's 1949 article was the first to criticize the theory in the Soviet literature, and seven more followed from there. Much of this work received the formal backing of the Soviet state, and all were approved for publication by the Scientific Council of the Institute of Organic Chemistry of the Academy of Sciences. Reviewing these articles today, it would appear that more than 400 Soviet scientists were in some way involved with the push to disprove resonance theory, their arguments centering on two primary themes: 1) it was not ideologically congruent with Soviet ideology, and 2) it did not recognize the work of previous Soviet or Russian scientists.

The framework established by the seven articles set the stage for the next chapter in the controversy: the approval of an official stance on resonance by the Soviet government. As part of the tradition of Lysenkoist thinking, Soviet scientists periodically held conferences where theoretical doctrine was debated and a specific piece of ideology was declared the "winner." These victorious ideas would, in turn, become the Soviet government's supported view, at which point no other competing theory would be allowed. In 1951 it was resonance theory's turn to be debated and, to the surprise of few, it was ultimately decided that Pauling's ideas were not to be considered as part of the Soviet doctrine surrounding the structure of molecules. Furthermore, according to a U.S. State Department document, a decree was issued that "all remnants of the mistaken conceptions of resonance" were to be eradicated.

The official Soviet position on resonance theory was disseminated in an article by O.A. Reutov in the *Journal of General Chemistry of the USSR* and titled, "Some Problems of the Theory of Organic Chemistry." In it, Reutov systematically attacked Pauling's ideas as "bourgeois science" and "alien reactionary ideas" that did not take into account Soviet foundational work on molecules.

According to Reutov, the "true" chemical structure of the molecule had been developed by Russian scientist Aleksandr Mikhailovich Butlerov, to whom Pauling had expressed "insufficient appreciation as well as an incomplete understanding and a perversion of the Butlerov theory of structure." So egregious were Pauling's omissions that Reutov was forced to conclude that the theory of resonance had been formulated, at least in part, for "the belittlement of the importance of Russian science."

Reutov's rhetoric only escalated from there. "[I]t is urgently necessary," he wrote, [to criticize resonance on the] "basis of Marxist-Leninist methodology." (A passage that all but admits that the controversy had little to do with scientific thinking.) This needed criticism would reliably be carried out by Soviet organic chemists who, to use Reutov's words, "form an army of many thousands strong." By the end of the article, it was clear: the Soviet scientific community was ready to attack Western ideas on the chemical bond, and Pauling was their primary target.

Because Reutov's article was originally published in Russian and chiefly circulated within the Eastern Bloc, Pauling did not find out about it for several months. In August 1951 he finally received a copy from the Consultants Bureau, a translation center specializing in scientific journal articles and based in New York. Pauling was understandably angered by the content and tone of the article, and expressed hope that "some action can be taken to stop this attack," which to him was a "very vigorous one."

*

Not long after learning about Reutov's publication, Pauling wrote a charged letter to the Soviet Academy of Sciences condemning the piece and upholding the importance of resonance theory as key to growth in the field of structural chemistry. In it, Pauling made clear his belief that the critical rhetoric was falsely placed and, more broadly, that "the attempted suppression of a part of science is based upon a faulty conception of the nature of science." Pauling took particular umbrage with the ideological arguments against his theory, finding them baseless since they did not intersect at all with scientific principles. He also warned of the dangers of refusing to incorporate resonance into an entire nation's scientific vernacular, judging that "Any chemist in the modern world who attempts to carry on his work without making use of the theory of resonance [...] is seriously handicapped." At the conclusion of the letter, Pauling took pains to praise the two Soviet scientists, Syrkin and Diatkina, who had translated his work, spread word of his theory among their colleagues, and ultimately been forced to recant their support. In Pauling's view, the duo were "among the most able chemists in the USSR."

Pauling's reply was met with scorn in the Soviet Union and, as a direct result, the Soviet Academy organized a special meeting to specifically discuss and condemn resonance theory. Held at the end of 1951, the meeting was attended by over 400 people and featured public denunciations of Pauling's work by no fewer than 44 prominent Soviet scientists. One especially notable speaker was Syrkin who, in his talk, began by suggesting that Pauling's "substitution of the real molecule by a set of resonance structures for the sake of convenience led to an arbitrary and speculative element, and brought about what looks like an explanation, instead of an actual critical analysis and discussion of the mechanism of chemical reactions."

While it is unclear if this notion represented Syrkin's true feelings, it is reasonable to presume that his words echoed the sentiments of many in attendance. As he progressed through his talk, Syrkin put forth a standard criticism of resonance, objecting to the notion that a molecule might not have a discrete structure. This argument, of course, was grounded fully in Soviet ideology, and particularly the belief that everything must be real and known.

As it turned out, Pauling's letter and the Soviet Academy meeting that followed marked the crescendo of the resonance theory controversy. Following the conclusion of the 1951 conference the debate continued to simmer, with small attacks volleyed by both sides for another two years. But in 1953 the vitriol quickly waned, specifically because of the death of Joseph Stalin. The end of Stalin's reign ushered in a new era for the Soviets, one where it was no longer imperative to uphold orthodoxy for fear of severe punishment. And the fact that the Soviet scientific community did not pursue its condemnation of resonance theory with any vigor following Stalin's demise suggests that the debate was never actually rooted in beliefs about problems with the science itself. More likely, resonance was just one of many convenient targets for Stalin and other Soviet figureheads in their war against ideas that were ideologically incongruent with Soviet doctrine.

*

Post-Stalin, the de-escalation of the Soviet resonance controversy continued to play out for at least another ten years, a time period during which the combatants' points of emphasis gradually shifted. Instead of arguing against the theory because it opposed Soviet ideology, Pauling's critics instead focused more on his failure to acknowledge the work of Russian scientists whom they argued were influential to his breakthrough. This difference marked a monumental deviation from the earlier platform for opposition. Before, the Soviets put forth a blanket rejection of resonance as an idea, and thus abandoned its use in developing their scientific work. Now, the Soviets tacitly began to recognize the utility of resonance while still arguing for credit that they felt was owed to luminaries of their past.

Pauling, in turn, also began to soften his rhetoric in defending the theory, perhaps in part because of the complex nature of his relationship with the Soviet Union. The years of the resonance controversy coincided with stateside accusations that Pauling was a Communist and a Soviet sympathizer (see Chapter 11). During this period, Pauling had also been very public in urging the Soviets to slow down the pace of their atomic weapons development and testing. In the midst of these heady issues, the battle to defend resonance was perhaps not quite so pressing, and it was in this context that the final phase of the resonance theory controversy took place.

The last chapter of the dispute began when Pauling gave a speech on resonance in Moscow, a talk that would have been unimaginable during Stalin's lifetime. Delivered on December 3, 1961, Pauling made his remarks to the chemistry division of the Soviet Academy of Sciences. At his hosts' urging, Pauling began by praising a Russian chemist, Mikhail Vasilyevich Lomonosov (1711–1765), for first theorizing that the properties of chemicals are a result of "their structure as aggregates of atoms."

Pauling's offering was not without calculation however, in that he also argued that Lomonsov was a free thinker who had used his "imaginative effort" and "high originality" to develop his ideas. Perhaps most importantly, Pauling stressed that Lomonosov's ideas were based on conceptual findings, not ideological doctrine. In this, Pauling was clearly trying to emphasize the ways in which Russian science had differed from the more contemporary Soviet approach, with old masters like Lomonosov developing their thinking on the basis of experimental data. By using Lomonosov as an anchor for this argument, Pauling made clear his continuing objection to the Soviet ideological attack on his theory.

The controversy had been brewing for well over a decade by the time Pauling gave his speech, so it would stand to reason that many in attendance still doubted the validity of resonance theory. And indeed, handwritten questions offered to Pauling at the conclusion of his lecture ranged from outright denial of resonance theory to asking whether or not MO theory or other explanations could instead be used to understand chemical structures and their properties.

Generally speaking though, these audience-generated questions were not ideologically based, perhaps because Pauling had deliberately chosen at the outset to "ignore the criticism [of his theory] on ideological grounds" and instead devoted most of his speech to systematically arguing in support of the science. The case that he made was clear and straightforward, but also imbued with undertones of sadness. In one instance, Pauling mourned the loss of scientific contributions that could have been made had the Soviet establishment bought in to resonance from the beginning. This sentiment came across most acutely in Pauling's lament that there was not a single "textbook of chemistry published by any Soviet scientist during the past ten years in which the theory of resonance is presented in a reasonably satisfactory way," an omission that would guarantee that future Soviet students would be "seriously hampered."

Publications authored in the months following Pauling's visit increasingly came to show that the Soviets no longer rejected resonance, but did want to be included in the story of its discovery. One notable example was a May 1962 article by G.M. Bykov titled "The Origin of the Theory of Chemical Structure." The piece was written in English, a clear indication of its intended audience, and argued that all chemical structure theory should be based on the work of Russian scientists. Though Bykov was willing to concede that Russian scientists such as A.M. Butlerov "did not develop and did not always support" other ideas about chemical structures, he was obviously "the founder of the theory of chemical structures" and a forefather of "the establishment of a correct conception of chemical atoms."

Responses of this sort allowed the Soviet academy to protect national pride while quietly updating their practices to align with the rest of the world. The shift in rhetoric also helped to end the extreme animosity that had been directed towards Pauling and, by the 1970s, the Soviets had essentially retracted any ideological objection to resonance theory. In fact, in 1972 the popular Soviet science magazine *Znanie Sila* (*Knowledge is Power*) published a retrospective titled "In the memory of a theory" that used interviews with M.E. Diatkina to recount the resonance story and its importance in the world of chemical structures.

*

Even though the resonance controversy was largely contested between Soviet scientists and Linus Pauling, many outside members of the scientific public, as well as the media, saw fit to weigh in on the dispute.

By and large, the reaction was quite supportive of Pauling, with most commentators coming to his defense. One scientist who definitely fit this bill was G.W. Wheland, who worked with Pauling to develop the theory. In a letter to the editor of *Chemical and Engineering News* dated August 4, 1952, Wheland denounced the Soviet attack on resonance as being driven by "patriotic, political, and ideological invective, which from the Western viewpoint, has no scientific content."

Chemist Moyer Hunsberger felt similarly, and in 1952 he delivered a series of lectures meant to thwart "the Russian chauvinistic ideological approach" to scientific inquiry. In them, Hunsberger labeled the Soviet position and its steadfast support of A.M. Butlerov an "extremely obvious exaggeration of [his] contributions to organic chemistry." Likewise, "the intensity and crudeness" of the Soviet condemnation of Pauling's ideas "appear[ed] to be without parallel in the annals of chemistry."

Other scientists added their voices in concert. Notably, Harvard University president James Bryant Conant argued,

> "If the Russians continue to attempt to force science to follow along a path determined by politics, Russia is sure to grow weaker. If Russians are not allowed to use the resonance theory or are deprived of scientific freedom in any other direction, Russian science will fall behind Western science and Russian technology will suffer. I think that Syrkin and Diatkina and other Russian chemists who have been criticized for using the theory are among the most able in Russia today. Their book, *The Chemical Bond and the Structure of Molecules*, is an excellent work. It is based almost entirely on resonance theory — and I think there is no substitution for that."

Pauling's scientific peers weren't alone in coming to his defense. Many media outlets also published articles about the controversy and, as with the academics, they were strongly in support of Pauling and his work. Perhaps the first mention of the controversy in the American press appeared in a July 15, 1951 *New York Times* article titled, "Soviets Dispute a Chemical Theory." In it, the author impugns the Soviets for not acknowledging the centrality of resonance theory and even asserts that the Soviet atomic bomb projects were necessarily informed by the work. As such, it was clearly hypocritical for the theory to be described as "bourgeois mysticism" when it had been so important to advancing goals fundamental to the national interest.

Later that summer, the California Institute of Technology News Bureau released a bulletin about the controversy that sought to summarize the Soviet attack and defend the importance of Pauling's ideas. "The theory is now an accepted part of chemistry," the News Bureau wrote, "at least in the Western world. It is taught in general and organic chemistry textbooks and has contributed to the simplification of the science."

These early analyses in hand, other newspapers began to report on the controversy, with some even drawing comparisons between the Soviets' stance and actions taken by the U.S. government a year prior. In particular, a *Seattle Times* piece from September 2 compared the dispute to accusations made by the House Un-American Activities Commission in 1950, wherein Pauling was branded a Communist. By recognizing and naming the similarities between the two controversies, the author of the piece put forth both an exoneration of Pauling's political activities in the U.S. as well as the perception of his science in the USSR. A steady flow of support continued from there, with the *Los Angeles Times*, *Newsweek*, *Time*, and the *Pasadena Independent* among those publishing popular articles that upheld resonance theory and denounced the Soviet position.

In 1952 the *New York Times* published a second article investigating the topic. This piece sought to explain resonance theory in basic terms that "every schoolboy" could understand while also responding to lingering objections from Soviet scientists. According to the article, scientists in the USSR had recently reaffirmed their denouncement of resonance theory, calling its ideas "mere illusions, senseless structures." These critics continued that "Pauling and his like are said to cherish perverted concepts which are typical examples of bourgeois thinking. The man to follow is A.M. Butlerov, who has a more materialistic concept of chemical structures." This established, the *Times* article then put forth a clarification. "It may be that resonance structures are illusory," the author wrote, but "That is not the point. A theory is not a statement of absolute truth. It is an invention, a tool. [...] So far, the theory of resonance explains what cannot be explained by older theories of valence." The piece concluded of resonance theory, "Benighted Western chemists will continue to apply it."

But not all outside observers were entirely behind Pauling, and one figure in particular deserves mention. In 1977, two years after his passing, a biographical memoir was written of Robert Robinson, the former president of the British Royal Society. In this remembrance, it was revealed that Robinson did not at all agree with Pauling's ideas on resonance.

Robinson and Pauling had known one another for over two decades and were clearly friends. When Robinson received the Nobel Prize in Chemistry in 1947, Pauling wrote to convey his "heartiest congratulation," and at nearly the same time Robinson helped Pauling to secure an Eastman fellowship at Oxford University. When Robinson's wife died in 1954, Pauling reached out to express his "deep affection." Some years later, Robinson's daughter gave Pauling a portrait of her father from 1955, which Pauling kept on his desk as a reminder of the times they had shared together "sitting on stools at the little eating house after the Royal Society Meeting" as well as their "boat trip down the St. John's River."

But despite this close personal connection, Robinson differed on the issue of resonance theory. In the biographical memoir, he was quoted as having referred to the work as "very misleading" and an "unfortunate contribution" to science. Pauling was upset by this revelation, so much so that he wrote a response in which he argued that Robinson's beliefs were "based entirely on misunderstanding or incompleteness of knowledge of the nature and early history of the theory of resonance."

The late Robinson's critique stood as an outlier though and, upon final reflection, it appeared that Pauling's experience with the Soviets did not leave a lasting mark. In a Royal Society paper about resonance theory that he wrote in 1977, Pauling addressed the controversy just once, and in it he did not even specifically mention the Soviets. Rather, Pauling's text reported that the theory had been "rather strongly attacked... because of the failure of critics to understand it." And with that, he seems to have summarized his feelings about the Soviet resonance controversy in as succinct a manner as possible.

6 One Year as President of the American Chemical Society

By Madeleine Connolly

Linus Pauling was elected future president of the American Chemical Society (ACS) in December 1947. In that capacity, he served as the society's president-elect for 1948 — during which time he was living in the U.K. as a visiting professor at Oxford — and officially took up his post as president in 1949. He formally assumed office on January 1, 1949, with Ernest H. Volwiler taking his place as president-elect.

The news of Pauling's election as ACS president was widely publicized in early 1948, with one short announcement reading:

> **"Chemists' Chief** — Dr. Linus C. Pauling, chairman of the division of chemistry and chemical engineering at the California Institute of Technology, has been elected president of the American Chemical Society. One of the world's leading theoretical chemists, he was chosen in a national mail ballot of the society's 55,000 members."

This same blurb was used in multiple newspapers with only slight variations. *Chemical and Engineering News*, one of the ACS's signature publications, released a slightly longer announcement in 1949 describing Pauling's achievements in the field, including his academic positions past and present, and his laundry list of awards and honorary degrees.

News of his election was initially well-received by scientists both within the society and outside of it, and Pauling received letters of congratulation from many of his new constituents, eager to express their excitement at being led by a chemist of such high ability and international renown. However, as seemed to always be the case with Pauling, political stances quickly became a point of contention between himself and others in the ACS.

While many members wrote to Pauling expressing their joy at the election of a political progressive (one correspondent, Bernard L. Oser of Food Research Laboratories, Inc., wrote that "The ACS has done honor unto itself by handing the gavel to a great liberal as well as a great chemist"), a larger and more vocal group of society members quickly withdrew their support.

In November 1948, near the end of Pauling's stint as president-elect, the first of the political controversies arose. In this instance, the catalyst was a pamphlet featuring Pauling's name at the top of a list of sponsors endorsing presidential candidate Henry Wallace. Wallace had served as vice president under Franklin D. Roosevelt and ran in the 1948 election as the Progressive Party nominee.

When he learned of the pamphlet, Henry C. Wing, the chairman of the ACS Board of Directors, wrote a letter to Pauling reprimanding him for publicly supporting Wallace, who had been accused of displaying Communist sympathies. In the letter, Wing suggested that "...your action has impaired the high position of our Society in the eyes of Congress and the nation, as it is impossible for you to separate your actions as an individual from that of the President of the Society." Wing then warned that "There can be no question concerning the loyalty to the United States of the vast majority of the members of the Society..." and notified Pauling that his behavior would be discussed at the next section meeting.

Other ACS members wrote to the board expressing similar concerns. One member, R.H. Sawyer, summed up the views of others in asserting that, although Pauling's name on the pamphlet was in no way overtly connected to the ACS (his professional associations were not listed), it still reflected badly on the society to have its members endorsing such figures. Sawyer then theorized that perhaps Pauling's name had been used without his consent and challenged him to confirm or deny his endorsement of Wallace. "Should [Pauling] fail or refuse to do so," Sawyer warned, "I call upon the American chemist to repudiate as a political crack-pot the man they have honored as a scientist by election to the office of President-Elect of the American Chemical Society."

A different ACS member, M.L. Crossley, offered a similar complaint, explaining that:

> "...I shudder to think of what some of the radio commentators may do with the information that the President of the American Chemical

Society is a sponsor of the Wallace-Marcantonio brand of politics. I register now and ever the strongest protest against any action by an officer of the American Chemical Society which may be used to convey the impression that I as a scientist and a member of the American Chemical Society subscribe to such a brand of dangerous, fanatical philosophy of government. I am a true American, believing in the principles of the philosophy of our democracy which has made this country the greatest land of freedom and opportunity the world has ever known."

While Sawyer and Crossley wrote to members of the ACS administration, others chose to address Pauling directly. These letters of criticism complained about his political views and expressed surprise that a man as intelligent as himself would be so liberal. The correspondents also requested that he release a statement clarifying his stances, and some even called for his resignation as president-elect.

Dr. Louise Kelly was one such author, lamenting to Pauling that despite "Having in the past had a high regard for your intellectual ability..." she was appalled to receive the pamphlet and find out that he was a Wallace supporter. Kelly continued, "I can understand why certain types of individuals, whose mental processes (I refuse to employ the word 'thinking') are as confused as those of Mr. Wallace, have been converted into followers of his, but I am at a loss to comprehend your position."

Pauling did not deign to accommodate requests for his resignation nor to issue a public statement on his views, but he did reply to the letters that were sent directly to him. Confident as ever, Pauling answered Dr. Kelly's letter by requesting that she re-evaluate her refusal to use the word "thinking" in conjunction with Wallace supporters, writing "I assume that you consider me to be a reasonably clear-headed fellow." He finished by assuring her that there was no need to be shocked and appalled — his political leanings had never been a secret, and he promised that "I haven't changed much in recent years, except that my health is not so good as it once was, my hair is getting thin on top, and my social conscience has grown a great deal."

Although the ACS was officially a non-political organization, the bulk of its membership seems to have been politically conservative, as is illustrated by the response to Pauling's endorsement of Henry Wallace. In reality, Wallace's platform was not Communist, but forward-thinking and focused on social justice. Wallace called Truman's government warmongering and hypocritical for pushing universal military training and a draft system, while veterans returned from war were too often rendered homeless and unemployed. He also believed that war profiteering on Wall Street was a source of the country's increasing militarism. Wallace likewise viewed the Marshall Plan as a tool for the U.S. to cement political and economic control of Western Europe; he encouraged aid for Europe but wanted to work through the United Nations to ensure that support was provided with "no political strings attached."

Wallace campaigned for world peace and civil rights for minority communities, and his platform included calls for desegregation and anti-lynching laws, outlawing the poll-tax, and providing federal inspections of polling places to ensure fair voting conditions. He also pushed for a fair employment practices act, desegregation of the government and armed forces, and the withdrawal of federal aid from any institution engaging in discriminatory practices. He advocated for the end of Jim Crow legislation and sought to eliminate discrimination against African Americans, Jews, and "citizens of foreign descent."

For his part, Pauling strongly believed that his actions as an individual remained separate from his role as ACS president, so long as he did not leverage his position to engender support for his political views. To this end, he made a point of not listing his academic or professional affiliations whenever he allowed his name to be used for political purposes. He also continually pointed out that he had never kept his views a secret and that his election had nothing to do with his politics.

In their exchange of letters, R.H. Sawyer retaliated that most of the society members who voted for Pauling would not have done so had they known of his political affiliations. Whether or not Sawyer was right, the

Wallace affair was the first taste of a continuing conflict between Pauling and the ACS that would taint his entire presidential year.

*

Having weathered the Henry Wallace controversy, Pauling formally entered into his presidential year and was immediately apprised of the need to address a significant problem. Namely, by the time he took office in 1949, the ACS had grown too large to function well from a financial standpoint. As a result, the society was constantly plagued by deficits on the order of several thousands of dollars, and was making budgetary adjustments left and right but still not breaking even.

One of Pauling's first obligations as ACS chief was to publish a President's Message in the January issue of *Chemical and Engineering News*. In his piece, titled "Our Job Ahead," Pauling directly addressed the society's financial difficulties, saying that he was sure the budget issues could be alleviated in a way that would support the society's cost of operations and the publication of its journals. Pauling also wisely sought to address "...a related problem, the discrepancy between the remuneration of members and the rising costs of living, and to help generally in improving the economic and professional status of chemists and chemical engineers."

From there, Pauling warned his membership against any failure to use science for good, emphasizing the society's responsibility to "...foster increased understanding and friendship among scientists of different nations" and reiterating the need to create a National Science Foundation. (Over the course of the year, Pauling addressed the issue of a National Science Foundation in several speeches to various ACS meetings, reiterating his support for the program, which had been proposed in the political arena but had not yet taken off.) The feedback to Pauling's column was generally positive, with many chemists taking especial note of Pauling's concern for their financial well-being.

In mid-January, Pauling began to promote another idea that he supported: the World Calendar. The World Calendar was a proposition that would regularize the lengths of months and theoretically be adopted by every country in the world. In Pauling's view, moving in this direction would rectify the "inconvenience" caused by the current calendar systems in use, and would make "every year the same." "The advantages," as Pauling put it, "are similar to those of an internationally accepted system of screw threads." At the time that Pauling expressed interest in the idea, it had already been endorsed by 17 nations around the world as well as multiple scientific organizations.

Correspondence exchanged between Pauling and other ACS executives indicates that there was a general willingness to bring up the idea at a meeting of the Board of Directors, and even some enthusiasm for its endorsement. The topic disappears from the record after January however, and obviously never went far in the national or international political spheres.

In February a mini-scandal of sorts hit the society when a collection of non-members protested the registration fees required of them to attend ACS conferences to which they had been invited to present. Pauling had not known that the society charged registration fees to invited non-members, and expressed vehement opposition to the practice when it was brought to his attention. But in this instance the president was overruled by executive secretary Alden Emery and others, who pointed out that the practice was necessitated by a clause in the society's constitution which would be difficult to change.

After a little more digging, Emery discovered that although the society as a whole struggled with funding — as did each local section and the society's various publications as well — all of the complaints regarding non-member registration fees came from the Division of Biological Chemistry. According to Emery, that particular division was, in reality, in "excellent financial condition." Pauling and Emery eventually concluded that the division's real issue was its lack of appeal to its target demographic. Fundamentally, the division was perceived by non-member biochemists to be more chemical than biological in focus, and therefore not an appropriate environment for sharing and publishing work that was essentially biological in nature.

As they puzzled over their continuing fiscal woes, the Board of Directors discussed the possibility of increasing membership dues to better support the society's many undertakings. This notion was countered by fears that increased fees might create a corresponding drop in membership that would defeat the purpose. In the end however, the fees went up.

At this same time, the February 1949 dismissal of Caltech graduate Ralph Spitzer from his faculty position in the chemistry department at Oregon State College (OSC) hit Pauling hard and brought his liberal political beliefs back into the limelight. Pauling was a mentor and friend to Spitzer and also an OSC alum; indeed, he had recommended Spitzer for the position at OSC. The grounds for Spitzer's dismissal were vague, but clearly political in their motivations. Then OSC president August Strand had accused Spitzer of harboring Communist sympathies based on his support of Henry Wallace in the presidential election and his advocacy of Trofim Lysenko's theory of the intergenerational inheritance of acquired characteristics, ideas that had originated in the Soviet Union and that differed from accepted Western theories. (Research in later years would ultimately prove Lysenko's theory to be incorrect, although it does bear some largely coincidental resemblance to modern epigenetics.) Importantly, rather than an endorsement of the theory, Spitzer's support was couched mainly in terms of defending the right for Lysenko's work to be read, respected, and considered through a scientific lens, and not discounted outright simply because of its Soviet origins.

As his standing at OSC dissolved, Spitzer wrote to Pauling to apprise him of the situation and to ask for his help in trying to get his job back. Pauling promptly wrote to President Strand to protest the removal of Spitzer on political grounds, which he considered to be an infringement of academic freedom since there had been no complaints of Spitzer's political leanings affecting his research or teaching.

The year before, Pauling had publicly spoken out against the House Committee on Un-American Activities for not giving scientists the chance to defend themselves against accusations of disloyalty. Now he found himself in the midst of a public feud with August Strand over Spitzer's dismissal, which culminated in a public rending of Pauling's previously strong bond with his undergraduate alma mater. Pauling and Spitzer later took part in a forum on the perceived incompatibility between academic freedom and Communism, joining a set of professors from other universities who found themselves in the same boat as Spitzer. Documentary evidence about how the ACS handled all of this is lacking, but the membership's broad reaction to Pauling's support for Henry Wallace leads one to suppose that the society was likely none too pleased about it.

*

Although Pauling's political activities were a source of irritation to many in the ACS, they did not seem to diminish his popularity as a public figure. During his year as president, Pauling traveled the country, speaking to a great many local ACS sections and receiving far more requests than he could possibly accept. Pauling's talks also routinely drew audiences that were much larger than many of the sections had seen before, and sometimes bigger than the sections had capacity to accommodate. After one such occasion, a regional section, in sending Pauling a note of thanks, even apologized to him in case "... those who had to stand became restless and in any way annoyed you."

Prior to a visit, Pauling generally offered each section a menu of three possible talks from which to choose: 1) "The Valence of Metals and the Structures of Intermetallic Compounds," 2) "The Structure of Antibodies and the Nature of Serological Reactions," or 3) "New Light on the Structure of Inorganic Complexes." As the months advanced, he introduced two additional possibilities: "Structural Chemistry of the Metallic State" and "Relations between Structural Chemistry and Analytical Chemistry."

Each stop on his speaking tours typically included a luncheon with a few of the section's higher-ups; a visit to a local university, factory, or laboratory where the majority of section members were employed; dinner out with members of the section; and finally the lecture at the end of the evening. For his talks, Pauling always requested a slide projector and a "good-sized blackboard," complaining rather bitterly whenever he arrived at a venue to find that the blackboard had been omitted or was not as large as he had anticipated. But sections were generally quite attentive to accommodating Pauling, with many changing the dates of their regular meetings and rearranging other speakers to suit the president's schedule.

In March Pauling attended the William H. Nichols Medal dinner, which was hosted by the New York section of the ACS. Pauling had himself received the Nichols Medal in 1941 for his work on the chemical bond. But for reasons that are unclear, the banquet at which the medal was customarily awarded had suffered a loss of prestige in the years since Pauling had received it. Hoping to restore the event to its former glory, the organizers made a concerted effort to invite as many high-profile ACS members as they could, and Pauling enthusiastically accepted the section's invitation to be present at the high table and to say a few words about the importance of the award. The 1949 recipient of the Nichols Medal was I.M. Kolthoff, now often cited as the father of analytical chemistry.

Pauling also served as the Edgar Fahs Smith Memorial lecturer for the Philadelphia section of the society. Smith was an internationally renowned chemist and educator as well as a past president of the ACS. Pauling's appearance drew a much larger crowd than the organizers had anticipated, filling the lecture hall completely including standing room, with many others turned away once the room reached capacity. The title of his lecture was "New Ideas on Inorganic Chemistry," focusing specifically on the structure of chemical bonds.

Pauling's travels were interrupted in early April 1949 when he was hospitalized for a "varicocele repair job." Though a minor operation, his recovery proved more difficult than the doctors had initially anticipated, and Pauling spent much of the next month bedridden. As a result, he was forced to cancel a speaking tour he had arranged on the eastern half of the U.S. as well as numerous other engagements that had been booked throughout the spring.

Forced out of the public eye for a moment, Pauling seems to have enjoyed a quiet period during which he went about his own work and attended to other presidential duties. These tasks included appointing various awards committee members as well as delegates to events and conferences that the ACS was invited to; participating in conversations related to budgets and logistics; and, once he was well, attending as many national-level meetings and local section events as he could.

Not all of Pauling's activities as president attracted widespread attention. In the middle of the year, the ACS forwarded a letter to Pauling from a John Albert, aspiring research chemist, who introduced a tale of job search woe as follows:

"Having been seeking employment for one year, having contacted more than one thousand prospective employers by resumes, having gone into debt more than $700 for bodily sustenance (food) during this period, having desired to marry, and having been refused even laboratory technician's employment..."

Albert included his résumé with the letter, requesting that Pauling critique it and offer any advice that he could to help a struggling chemist find work.

Pauling and ACS executive secretary Alden Emery both took note of Albert's pluck in contacting the president of the ACS as a job coach, and they decided to offer what help they could. After a little digging, Emery discovered that although Albert had spent five years studying full-time for his undergraduate degree, he never actually finished it, and from 1942 to 1949, a period of seven years, he had held six different jobs. In addition, the types of positions he was applying for — roles like lab supervisor — were too ambitious for someone lacking a degree and a strong record of long-term employment.

Pauling centered his response to Albert around two questions: why hadn't he finished his bachelor's degree? And why had he changed jobs so frequently? Albert replied that he had "not succeeded" in his chemistry courses at university but had earned A's in his music and German classes, and that, upon reflection, maybe he didn't possess the skillset necessary to do well in chemistry. This moment of introspection moved Albert to change career paths and seek employment in a different field.

Pauling responded one final time to congratulate Albert on making the effort to discover his true passion, assuring his correspondent that "...sooner or later you will be successful." Emery wrote back as well with the suggestion that he take an aptitude test that might help reveal professions to which he was well-suited. Albert promised to stop in at the office that Emery had recommended to inquire about tests of this sort. In doing so, he admitted that he was concerned about the fees associated with the process but would "...not close my eyes to any possibility which might uncover this enigma's secret..."

*

Meanwhile, several months had passed since Ralph Spitzer's dismissal from OSC, but Pauling had not forgotten the indignity. In June he took part in a luncheon discussion panel titled "Should Communist Party Membership be Grounds for Dismissal from a College Faculty?" In it, Pauling argued the negative, and Dr. George C.S. Benson, the president of Claremont Men's College, argued the positive.

The foundation of Benson's thinking was that the Communist party required its members to follow the party line in every aspect of their lives. As a consequence, adherent professors would be obligated to sneak Communist philosophies into their lectures for purposes of secretly indoctrinating students. Benson also held that a Communist party takeover would prevent minority groups from organizing, and that a country whose greatest legacy is liberty could not support the rise of a party which would suppress civil rights, violently overthrow the government, and engage in "intellectual double-dealing."

Pauling's counter was that "We are in danger, not from Communism, but from loss of principles." He pointed out that many political liberals and progressives, including himself, had been attacked for "Communism" though they were neither Communist party members nor sympathizers. Rather, the label had become an umbrella accusation used to derail or overpower any individual or group that did not conform to appropriately conservative political beliefs.

Pauling then proposed that establishing Communism as grounds for dismissal from a university faculty body showed disrespect for the academic integrity of professors and students alike, by implying that students could not be trusted to think for themselves and that allowing Communism a platform at all would be enough to ensure that it prevailed as a dominant ideology. Pauling concluded that it was cynics such as Benson, and not liberals, who ought to be brought for questioning before the Committee on Un-American Activities.

Though there is no documentary evidence for how the ACS reacted to Pauling's stance on academic freedom, it was not long before his continued political activities brought him back under scrutiny in a major way. In July Pauling received a letter from the Hanson, Lovett & Dale law firm, which had been hired by Dr. Roger Adams, chairman of the Board of Directors of the ACS. Through the lawyer, Adams complained that Pauling had utilized his title as society president in his endorsement of an ad that had appeared in the *New York Times* the previous day. The ad was titled "Tom Clark's Police State," and as usual was a political statement that ran counter to conservative sentiments. The lawyer, Elisha Hanson, informed Pauling that the use of his title in such a manner was unauthorized and a violation of society regulations.

Pauling wrote back three days later, stating that "I am very much troubled by the information...[that] the name of the American Chemical Society is used in connection with my name, in this political activity," and that he had asked specifically that his professional affiliations not be listed. Pauling also expressed his disappointment that a lawyer had been contacted about the matter before anyone from the society had bothered to ask him for a clarification. Though mostly a matter of miscommunication, the society's knee-jerk reaction to a relatively minor offense serves as evidence of the rift that had grown between Pauling and the ACS little more than halfway through his presidential year.

*

As Pauling moved into the second half of his term as president, the organization's financial issues briefly took center stage. In July a committee tasked with analyzing the society's expenditures and incomes began working in earnest. In particular, the committee sought to compare the costs associated with producing each of the society's various publications against the income generated by subscription rates and member dues.

Pauling's administrative records indicate that financing its publications was one of the ACS's greatest monetary hurdles. And while the issue was beyond the committee's ability to resolve, there was talk of increasing membership dues and publication subscription rates, decreasing financial support for certain niche publications in favor of more lucrative ones, and contacting chemistry-dependent industries that frequently took out mass subscriptions to ask for additional financial support. Ultimately it was decided that the 1950 membership dues would be increased; an explanation and defense of this move was published in the end-of-year report that was circulated to the full membership.

By late August, Pauling was once again attracting attention from his critics within the society. This time, Pauling's plans to travel to Mexico City to attend and present at the American Continental Congress for Peace were the source of the controversy. One especially prominent detractor, ACS board member A.C. Elm, wrote a letter to an unspecified recipient requesting that Pauling be dissuaded from attending the conference and, if that failed, that he be asked to resign from his leadership position. In his letter, Elm wrote that many ACS members were "greatly disturbed" to learn that Pauling was a sponsor of the meeting, which was "inspired and dominated by comunists [sic]."

Other members agreed that the act of requesting the president's resignation over his political activities was unprecedented but necessary in light of the potential smear on the society's reputation. Subsequent correspondence flew between ACS members, including every board member except for Pauling (Elm specifically stipulated as much in his first communication), trying to rally the organization against its

president and "...use their influence to prevent Pauling from embarrassing the Society."

These efforts failed and, in September, Pauling attended the Mexico City meeting. While there, he presented a talk titled "Man — An
Irrational Animal," in which he endorsed the idea of a world government, headed by the United Nations, to which all countries would need
to transfer their sovereignty. Pauling believed that such an arrangement would guarantee world peace by uniting the globe under a single
umbrella rather than a collection of competing nations.

To this end, Pauling placed the responsibility for seeking and
ensuring peace into the hands of citizens instead of governments, since
the latter's historical impulse toward national sovereignty had frequently
been antagonistic to the cause of peace. Pauling believed that international problems like hunger would never be solved while war continued
to waste resources and cause divisions between countries. He also took
particular aim at a segment of the scientific population for focusing on
weapon development projects instead of conducting work to better the
human condition.

Just two weeks after the American Continental Congress for
Peace, Pauling presided over the national meeting of the ACS. Held in
Atlantic City, New Jersey, the meeting was the largest in ACS history,
with a record 1,064 papers presented in 151 sessions over five days.
Such was the scale of the meeting that, in addition to space reserved
at six different hotels, three temporary rooms were constructed at the
main convention hall to satisfy meeting requirements. Several field trips
to glass, wine, and food producers were included on the conference
program as well.

Pauling gave his presidential address, "Chemistry and the World of
Today," on September 19, opening with a timely question:

"What can I say under the title 'Chemistry and the World of Today?'
My answer to this question is that I can say anything, discuss any

feature of modern life, because every aspect of the world today —
even politics and international relations — is affected by chemistry."

He went on to give examples of scientific achievements that had fundamentally altered Western society during the war years, including the
invention of nylon and the medical application of penicillin. He then argued
that the majority of scientific discoveries that are significant to modern life
come about as a result of basic rather than applied research, noting that it
is impossible to arrange or design life-changing discoveries. In emphasizing
this point, Pauling used the example that "Nobody, not even Einstein himself, could *plan* [Pauling's italics] to discover the theory of relativity."

From there, Pauling complained acridly about the hesitation of
various institutions to fund scientific equipment. In so doing, Pauling
quoted a Russian physicist P.L. Kapitza, who, in a 1943 speech before the
Soviet Academy of Science, had asked, "When you look at a painting of
Rembrandt, are you interested in the question of how much Rembrandt
paid for his brushes and canvas? Why, when you consider a scientific job,
do you want to know the cost of apparatus of the material used on it?"

Pauling was unequivocal in his belief that the benefits of scientific achievement far outweighed the costs involved in producing
the work, never mind the cost of equipment. As a means of easing
the burden on educational institutions, Pauling reiterated his support
for the creation of a National Science Foundation and stipulated that
funding should be made available to universities and research insti-
tutes without limitations on how it could be used. He also suggested
that corporations that rely on scientific discovery for their products
bore an obligation to fund a large chunk of that research by providing
similarly unrestricted research grants. Pauling felt that $250 million
a year in federal money through a National Science Foundation, and
$75 million a year from science-dependent industries, would work
well as starting levels of support. There were some concerns, raised by
Pauling himself, that such large subsidies might invite inappropriate
outside influence on scientific studies. But in general it was agreed

that the need for funding and the value of research were greater than the associated risks. Notably, Pauling also believed that the smaller contribution from industry would still be sufficient to protect against a government monopolization of science.

Newsweek's reporting on the annual meeting portrayed Pauling's address as the highlight of the convention, while making passing mention of an incident where "...a rare chemical element disappeared from the convention hall where it was on exhibit." The element was indeed very rare, and quite expensive as well. Called promethium (originally spelled prometheum), it had first been synthesized at the Oak Ridge National Laboratory during the war years and had been kept secret for reasons of national security. According to the article, the sample on display weighed only 2 milligrams yet was worth about $120,000 in 1949, which is equivalent to well over a million dollars today. After doing the math, the *Newsweek* reporter who filed the story concluded that

> "...the national debt of the United States would buy about 10 pounds of
> prometheum. The loss was deplorable but completely unpreventable,
> for prometheum is radioactive and slowly disintegrates."

The implications of the last sentence are never explained; instead, the article moves on to a discussion of new processes for electroplating aluminum coatings to other metals. Whether the promethium was stolen or somehow degraded down to nothing during the course of the conference is unclear, though theft seems more likely considering its monetary value (and its 17.7-year half-life).

One research paper presented at the national meeting caused turbulence for the ACS in the months that followed. In Atlantic City, Dr. Michael Somogyi gave a talk on the treatment of diabetes in which he stated that most cases could be managed through dietary means alone and did not require the use of supplementary insulin. In fact, Somogyi argued that doctors were overprescribing insulin treatments to the point of afflicting their patients with "insulin poisoning."

The American Diabetes Association (ADA) was infuriated by this stance, and complained that medically driven research with severe implications for the treatment of a disease should never have been presented by a chemist at a chemistry convention. The ADA also pointed out that, in allowing the paper to be presented, the ACS had undermined the ability of medical practitioners to effectively treat diabetics, since many patients had begun to request Somogyi's insulin-free treatment plan. The association further stressed that Somogyi's research had not been clinically validated and did not actually include a "treatment plan" per se, but rather consisted of a series of recommendations for treatment possibilities and a call for further research.

An ADA representative wrote to Pauling that Somogyi's actions had the potential to "jeopardize the lives of persons under treatment for diabetes" by encouraging potentially lethal abandonment of treatment plans involving insulin and weakening patients' trust in their doctors. The representative concluded that changes in the scientific understanding of medical treatments should not be released to the public until clinically approved by qualified medical practitioners, and then released by an appropriate scientific journal or medical body.

The Somogyi controversy was one of the last major topics that Pauling was forced to address during his term as president of the ACS and, as 1950 approached, he was ready to move on to other priorities. He was elected to the Executive Committee of the Board of Directors in 1950 and continued to serve the ACS in various capacities until 1963, when he resigned in protest of the society's muted reception of his Nobel Peace Prize. Then, as before, controversies surrounding his non-scientific work continued to hound his relationship with the nation's most prominent collection of chemistry professionals.

7 The Theory of Anesthesia

By Trevor Sandgathe

Linus Pauling and Frank Catchpool shaking hands at a Caltech function, 1963.

Until the 18th century, anesthetics were typically concocted from the local flora by herbalists and chemists. Opium, for example, is thought to be one of the oldest prepared anesthetics, distilled from poppy flowers farmed by Sumerians as early as 4000 BC. In the late 1760s, the great British scholar Joseph Priestley discovered the anesthetic power of nitrous oxide in its gaseous state, thus deprecating most conventional herbal anesthetics. Following Priestley's discovery, the international scientific community launched a number of small-scale investigations into potential anesthetics, eventually resulting in the medical use of ether, chloroform, and other gases. And in 1803, Friedrich Wilhelm Sertürner distilled morphine from pure opium, creating yet another wave of interest among researchers.

But despite these pronounced advances in practice, little was known about the properties of anesthetics, leading researchers to puzzle over a number of central questions. What caused the numbness and unconsciousness? Why were the effects of anesthesia reversible? What made some anesthetics more powerful than others? A few intrepid anesthesiologists suggested that anesthetic gases formed a sort of fog in the brain, or that they caused the nerves or brain matter itself to coagulate. Unfortunately, without access to advanced medical and chemical techniques, and lacking a sophisticated understanding of brain functioning, scientists harbored little hope of uncovering the precise mechanisms behind anesthesia.

In 1847 the German polymath Ernst von Bibra decided to tackle the problem. In years prior, von Bibra had specialized in the study of intoxicants and poisonous plants and, as a result, had accumulated a great deal of experience with the various medicinal compounds derived from flora. Von Bibra's idea was that anesthetics might dissolve fats in human brain cells, resulting in a temporary loss of consciousness and normal brain activity. He further theorized that, at some point after the anesthetized state had been induced, the anesthetic would eventually cycle out of the brain, thus permitting the brain's cells to steadily return to their natural rate of functioning.

Von Bibra realized that, if true, his theory would explain the temporary yet reversible unconsciousness induced by anesthesia and, in the process, revolutionize the scientific understanding of how the brain works. As it turned out his research was largely ignored for a half-century but, in the late 1800s, von Bibra's theory resurfaced and attracted the attention of several researchers who would go on to revolutionize the study of brain chemistry.

*

In 1896, Hans Horst Meyer, a German pharmacologist and Director of the Pharmacological Institute at the University of Marburg, became interested in von Bibra's work. Meyer hypothesized that anesthetics were hydrophobic (repelled by water) and in turn were attracted to other hydrophobic molecules. Lipids, the fatty molecules in brain cells, are also hydrophobic, as evidenced by the separation of lipid-based substances (such as vegetable oil, grease, and butter) in water. Meyer believed that this mutual hydrophobia led anesthetics to bond to and dissolve the lipid molecules in brain cells. He further argued that increasingly hydrophobic anesthetic molecules were capable of more readily forming stronger bonds with lipids, thereby increasing the potency of the anesthetic effect.

To test his hypothesis, Meyer began a small-scale research program on anesthetics, using his university standing to acquire the necessary assistants and apparatus for his experiments. His intention was to demonstrate some degree of correlation between a substance's ability to bond with fatty substances and the strength of its anesthetic power. As a means of assessing interactions between anesthetics and lipids, Meyer measured the solubility of known anesthetics (including, but not limited to, ketones, alcohols and ethers) in olive oil, which was meant to represent the fatty molecules in brain cells. He then tested the same anesthetics on tadpoles, measuring the quantity of anesthetic agent required to induce what he defined as "abnormal behavior." Though his use of tadpoles as experimental subjects led to imprecise and often subjective observations, he was able to positively correlate lipid solubility with anesthetic potency. In doing so, Meyer was able to offer experimental support for von Bibra's hypothesis, which he published in 1899.

In 1901, Charles Ernest Overton independently published his own theory of anesthesia. In addition to confirming Meyer's findings on lipid solubility, Overton also discovered that the power of an anesthetic was unrelated to the method by which it had been delivered. In other words, lipids in the brain were affected by anesthetic agents regardless of whether they had been administered in a liquid or gaseous form.

Because Meyer and Overton, both established researchers, came to the same conclusions using different experimental methods, their work gained traction in the scientific world and quickly became known as the Meyer-Overton theory of anesthesia. In its simplest form, the theory claims that once an anesthetic agent reaches a critical level in a lipid layer, the anesthesia molecules bond to target sites — sometimes called receptors — on the lipid molecules, in the process dissolving the fatty part of the brain cells affected by the anesthetic agent. In response to the dissolution of the lipid layer, the brain reaches an anesthetized state and the patient is rendered unconscious.

For many in the early 20th-century scientific community, the theory seemed to adequately describe the well-established relationship between anesthetics and lipid solubility. The ideas had been substantiated by multiple experimental tests and, in the end, served as the best existing explanation of the phenomenon. Meyer and Overton seemed to have decoded the mystery behind a major medical practice.

As is common with big discoveries however, the Meyer-Overton theory eventually succumbed to further scrutiny. Nearly six decades after the duo's original publications, researchers were finally able to identify a key flaw in the lipid theory: namely that anesthetics interacted with lipid-free proteins in the same way that they interacted with lipids. This suggested that anesthetics did not require lipid target sites for binding, but could instead bind to other sites with the same resulting anesthetic effect. This discovery greatly reduced the perceived importance of lipids in the anesthesia-brain interaction.

Researchers also found that, as anesthetics in a given series of tests became increasingly hydrophobic (through the lengthening of the carbon chain), their potency did not increase indefinitely. Instead, molecules appeared to reach what is known as a "cutoff point" where otherwise-effective anesthetics lose their ability to sedate the brain. According to the Meyer-Overton theory, this loss of anesthetic effect would imply an inability to bond with lipids. Scientists, however, found that long-chain anesthetics continued to bond with lipids despite the loss of anesthetic ability, further strengthening the argument that anesthetic-lipid bonds are not responsible for the sensory-altering effects of anesthesia. With these breakthroughs, the Meyer-Overton theory was crushed. If anesthetics could be effective without bonding to lipids, and could also be ineffective *when* bonded to lipids, the original Meyer-Overton theory could no longer be considered valid.

*

In part because of his love for puzzles and conundrums, Linus Pauling was always on the lookout for new and difficult projects, and by 1952 he was among those interested in trying to better understand anesthesia.

During the late 1940s and early 1950s, Pauling served as one of 12 scientists on the Scientific Advisory Board for Massachusetts General Hospital. In December 1951, he traveled to Boston to attend an advisory board meeting during which Henry K. Beecher, an anesthesiologist later known for his work in medical ethics, gave a talk on xenon as an anesthetic. Pauling was baffled by Beecher's findings because he knew that xenon, due to its full electron shell, is highly unreactive. According to conventional logic then, xenon should have had virtually no biological effect.

For several weeks following the conference, Pauling thought about the problem in earnest, but simply could not tease out the answer with the little bits of information that he had on hand. In 1952 Pauling became interested in methane hydrates and began a small-scale research program to study the properties of related compounds. He assigned Dick Marsh, a graduate student at Caltech, to the project of manufacturing and studying chlorine hydrates. By combining chlorine with chilled water, Marsh was able to create the compounds, which he then examined using X-rays.

The results were interesting. Marsh's pictures seemed to indicate that the chlorine molecules were forming an ice-like tetrahedral cage around the water molecules, effectively trapping and freezing the entire unit. After studying the data, Pauling realized that, like chlorine, xenon was capable of forming hydrates. From there he proposed that, if xenon hydrates were created in the brain, they would block the flow of ions through their lipid channels, essentially freezing all communication in the brain and rendering the subject unconscious. As brain tissue itself is approximately 78% water, there is more than enough liquid available to allow for hydrate formation. Pauling estimated that as little as 10% of the water in the brain would need to be incorporated into hydrate molecules to result in insensitivity to pain and unconsciousness.

As promising as this hypothesis seemed, it possessed one glaring flaw: A xenon hydrate becomes unstable and deteriorates at only two or three degrees above the freezing point of water. The human body's native temperature is therefore approximately three times that necessary to decompose xenon hydrates. Because of this, Pauling realized that hydrates couldn't possibly explain xenon's strange effect on the body, and he was forced to accept that, without undertaking his own research program on noble gases, he would be unable to develop a solution to the xenon predicament. He laid the problem aside, assuring himself that he would return to it in due time.

In 1954 Pauling was awarded the Nobel Prize in Chemistry and his life became a whirlwind of activity. Very soon he became a staple on the university lecture circuit, gave scores of interviews, and began applying his new-found fame to the peace movement. What little time he had left was spent supervising graduate students and applying for grants at Caltech, leaving scant opportunity for new research projects.

Nevertheless, the xenon question was not forgotten. In 1957 Pauling gave three talks on the chemical bond that were filmed by the National Science Foundation and distributed to institutions around the country. In his second lecture, Pauling enumerated a revision of his 1952 theory on xenon hydrates, suggesting that they might be stable up to ten degrees above the freezing temperature of water. Even still, the revision wasn't enough to make hydrates viable at body temperature. What Pauling needed was a breakthrough, something that would fundamentally change how he thought about the hydrate-temperature interaction.

By his own recollection, that breakthrough came in April 1959 while he was reading a paper on alkylammonium salt, a crystalline hydrate resembling the protein side chains found in the brain. The paper claimed that alkylammonium salt, which is also a clathrate similar to the xenon hydrates, was stable up to 25°C (77°F). Pauling realized that the dodecahedral chambers contained within the alkylammonium hydrate structure were strikingly similar to those formed in xenon hydrates. He hypothesized that xenon atoms introduced into the bloodstream could become trapped in the alkylammonium hydrate, thereby stabilizing the structure and raising its heat tolerance to approximately 37°C (98.6°F), thus preventing the hydrate from decomposing at body temperature.

Pauling now believed that, once the alkylammonium hydrate crystals had formed with xenon, they would prevent normal electrical oscillations and block the flow of ions in the brain, inducing anesthesia. Furthermore, the hydrates would gradually dissipate, in the process allowing the anesthetized brain to resume normal functioning. In short, Pauling had found the key to a new, seemingly workable model which would soon be referred to as the "Hydrate Microcrystal Theory of Anesthesia."

*

After nearly a decade of puzzling over the mechanisms of anesthesia, it looked as though Pauling had solved a problem that had baffled scientists for more than a century. But in order to prove the theory, he needed to begin the experimentation process. And for that, he needed a lead researcher.

In the summer of 1959, Linus and Ava Helen traveled to central Africa to visit Albert Schweitzer's famous medical compound in Lamberéné. There, they met Frank Catchpool, Schweitzer's chief medical officer. Pauling found Catchpool to be both intelligent and engaging; the two men spent a great deal of time together, touring the compound and discussing a variety of medical and scientific problems. Thoroughly impressed by the young physician, Pauling suggested that he apply for a position at Caltech. Shortly thereafter, in 1960, Catchpool became a researcher in the chemistry division under Pauling's direction.

Once Catchpool had arrival in Pasadena, the two men discussed the problem of anesthesia. As they talked, Pauling began to formulate experiments for the new researcher to conduct. Before long, Catchpool and his assistants were hard at work attempting to verify Pauling's theories. Success would not come easily, however — try as he might, Catchpool could not find a definitive link between microcrystals and anesthesia. In a June 1960 letter to his son, Peter, Linus described the experimental anesthesia work in which he and Catchpool were engaged. In it, he confessed that "Dr. Catchpool is just beginning a series of experiments on the effect anesthetic agents have in changing the brain waves of an artificial brain, made out of gelatin. I don't know whether anything will come of this or not. I like the whole theory of anesthesia, but it is hard to think of good experiments to carry out in connection with it."

Not long after, Pauling and Catchpool shifted their tactics. Instead of trying to demonstrate the anesthetic effects directly, the team opted to approach the problem tangentially. Rather than proving that hydrates were responsible for the anesthetic effect, they would try to show that lower body temperatures (which would increase hydrate formation) would allow known anesthetics to act more quickly and with a stronger effect. From there, they might correlate high rates of hydrate formation with an increased anesthetic effect.

To test this approach, Catchpool and his assistants brought dozens of goldfish to the lab, each in its own temperature-regulated bowl. There, they mixed various anesthetic agents into the bowls, hoping to find that the fish kept in lower-temperature water would become more quickly anesthetized than those in warmer water. Unfortunately for the researchers, goldfish proved to be difficult test subjects. Much like Hans Horst Meyer's tadpoles some 60 years before, the Catchpool group's fish were almost impossible to observe objectively, and the experiment quickly devolved into a guessing game. To make matters worse, Pauling's Caltech colleagues were beginning to take notice of his strange experiments, leading to more than a few raised eyebrows.

But Pauling was unwilling to admit defeat. He felt strongly that his theory had merit and was determined to publish it before another researcher had the chance. After a few preliminary lectures on the subject in early 1960, Pauling decided that he was ready to unveil the concept to a larger audience — with or without experimental evidence. He spent parts of the spring and summer working on the paper, alternating between his office at Caltech and his home near Big Sur. A year later, in July 1961, Pauling published "A Molecular Theory of General Anesthesia" in *Science* magazine.

Pauling and his team thought the paper would make a major splash in the medical world. As the first viable theory of anesthesia in decades, they expected chemists, biologists, and practitioners to be clamoring for details about his findings. Instead, the response was muted. A few anesthesiologists took note, but the scientific community as a whole remained unaffected.

To make matters worse, a rival paper on anesthesia was published in the *Proceedings of the National Academy of Sciences* that same month. The competing paper, published by Stanley L. Miller, a researcher at the University of California at San Diego, put forth an idea that was similar to Pauling's. Miller's piece claimed that tiny "icebergs" formed around the gaseous anesthetic agents, preventing normal electrical oscillations and the flow of ions. And because Pauling's paper had been published just prior to his own, Miller also had a chance to address Pauling's theory. In a "note added in proof," Miller wrote, "Since this article was submitted, a paper by L. Pauling has appeared (*Science*, 134, 15 (1961)) in which a similar theory is presented. Pauling proposes that microcrystals of hydrate are formed during anesthesia, these crystals being stabilized by side chains of proteins. In spite of any possible stabilization of hydrate crystals by protein side chains, it appears doubtful that crystals could be formed. The gas-filled 'icebergs' could be considered equivalent to Pauling's microcrystals, except that the 'icebergs' are much smaller and are not crystals in the usual sense."

Thus rebuked, things were looking gloomy for Pauling. It was bad enough that his theory had gone largely unnoticed, but with Miller's idea being so similar to his own, and published so closely to it, his own work no longer looked entirely original. Over the next 18 months, Pauling did his best to promote his theory, giving a few speeches on the work and even trying to draw attention to the similarities between his and Miller's publications, in hopes of gaining credibility. Unfortunately, the scientific community simply wasn't interested.

It is difficult to unpack the exact reasons why Pauling's theory was so effectively ignored. After all, he was a Nobel Laureate, a prominent member of the international scientific community, and a well-known public figure. Moreover, he was presenting a novel solution to a problem that had troubled scientists since the mid-1800s. Given the time period though, and events in Pauling's personal life, it is possible to imagine some of the contributing factors. First, one must consider the impact of his political activities. Not only had Pauling sacrificed huge amounts of his laboratory time to lectures and peace demonstrations, he had also attracted the attention of the Senate Internal Security Subcommittee, a body designed to seek out and interrogate suspected Communist sympathizers. The Senate committee hearings, public appearances, and meetings with lawyers ate up much of his time during the first part of the decade, leaving Pauling with little room to further research or promote his theory.

Pauling was also at odds with Caltech administrators during the early 1960s, a relationship frayed by his progressive political activities and, to a lesser degree, his unconventional research projects. Without a strong base of administrative support, it had become more difficult for him to access personnel and lab space, conduct research, and publicize his findings. This break between Pauling and the staff would culminate in his 1963 resignation from Caltech and subsequent move to the Center for the Study of Democratic Institutions.

Last, and perhaps most important, was Pauling's research philosophy. Pauling believed in what he called the "stochastic method." In principle, the stochastic method requires an individual to apply his or her knowledge of a given subject to a particular phenomenon with the intention of developing a hypothesis regarding the phenomenon, absent much (or any) unique laboratory data, which might be generated later. In layman's terms, one might refer to the process as making an educated guess and then designing experiments to see if the guess is correct.

But to suggest that Pauling simply guessed is an overstatement. Rather, he combined the available information about a subject with his considerable skill as a scientist to formulate what he saw as a viable, working theory. Then he would usually hand his findings off to other researchers, leaving them to do the experimental work. In most cases, the arrangement worked well. While he was primarily interested in building theoretical foundations and less inclined toward the tedious work of running experiments, most others lacked Pauling's creative genius, and felt more comfortable in the structured, hands-on laboratory setting. Normally this resulted in a sort of symbiotic relationship in the Caltech laboratories, but it also meant that not all of Pauling's theories received full attention. If no one chose to work with Pauling's theories, or if the research methods deployed proved unsuccessful, the theory was sometimes destined to gather dust in one of the Institute's filing cabinets. It's at least possible that the difficulty of conducting appropriate experiments had a hand in silencing Pauling's hydrate microcrystal theory.

Whatever the reason, Pauling's theory now stands as little more than a footnote in the history of anesthesiology. After its publication in

1961, it quickly faded out of the picture and the field was, yet again, left without a single agreed-upon theory.

*

While Pauling still supported his own ideas, his fellow scientists remained uninterested and he gradually disappeared from the scene altogether. Fortunately, the problem was not forgotten for long. Beginning in the mid-1970s, Nicholas P. Franks and William R. Lieb, researchers at the Imperial College in London, began to develop a new theory. They suggested that anesthetics are, in fact, similar to conventional pharmaceuticals, and that anesthetic molecules are able to bond to protein receptors in the brain where they manipulate specific ion channels. Like the hydrate theory, the protein theory posits that, by affecting the brain's ion channels, the anesthetics would have the ability to disrupt brain functions and result in unconsciousness.

Franks and Lieb spent several years testing their hypothesis by measuring the effects of various liquid anesthetics on isolated, lab-grown proteins. In 1984 they published an article in the journal *Anesthesia* titled "Seeing the Light: Protein Theories of General Anesthesia," which introduced their work to a wider audience and suggested that, through extensive testing, scientists might be able to identify the correlations between specific anesthetics and binding sites. This, in turn, would allow researchers to predict the effects of a given anesthetic and eventually develop improved synthetic chemicals.

In order to positively demonstrate the relationship between anesthetics and protein receptors, researchers in the United States and Switzerland began developing genetically modified mice. These test subjects, known as knockin mice, lacked specific proteins thought to be affected by a given anesthetic. By using an anesthetic on a supposedly immune mouse, the researchers were able to pinpoint correlations between anesthetics and proteins. With improved technology, the researchers were eventually able to minimize the necessary genetic changes by altering amino acids within the proteins. This allowed the researchers to avoid eliminating macromolecules within the knockin mice, resulting in a more authentic testing process.

The results from the knockin mice experiments proved monumental. Through extensive testing, researchers were able to locate and identify specific interactions between anesthetics and protein receptors, meaning that, for the first time in over a century of studying anesthesia, scientists were finally able support theoretical claims with conclusive experimental data.

Unfortunately, this breakthrough did not solve the mystery completely. Scientists also have discovered that anesthetics in gaseous form, which are commonly used to induce general anesthesia, do not necessarily adhere to the same principles as injected anesthetics. Notably, inhaled anesthetics do not seem to bind as tightly as their injected counterparts, and instead pass over a huge number of receptors rather than triggering a single one. Though a great deal of disagreement exists among scientists, it is widely believed that gaseous anesthetics affect anywhere from three or four types of receptors to over 100. To further complicate this issue, there is disagreement on the question of whether every receptor affected by the gas contributes to the anesthetic effect.

From Ernst von Bibra to Pauling to Franks and Lieb, the theory of anesthesia has had a bumpy ride. But, with each researcher and each breakthrough, we have moved a little closer to a better understanding of our biological selves. With a little luck and a lot of hard work, the years to come will yield even more progress and, undoubtedly, more questions.

8 Lloyd Jeffress

By Trevor Sandgathe

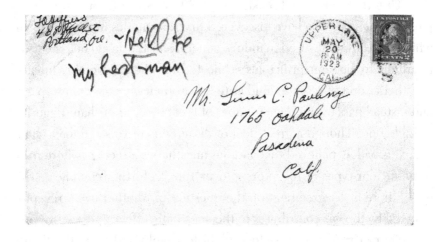

As a child, Linus Pauling had relatively few friends. After moving from Condon, Oregon to Portland, the death of his father and subsequent poverty forced him to work when not in school. The remainder of his time was consumed with studying and household chores, leaving little room for companionship. The young Pauling was also unusually introspective and self-reliant, capable of quietly entertaining himself without supervision. Nevertheless, even the busiest and most independent children need friends.

In 1913, while walking home from school, Pauling began talking with another young boy, Lloyd Jeffress. The two quickly discovered a mutual interest in science and natural phenomena, and Lloyd invited Linus to his home to view a chemistry experiment. Pauling readily agreed and, within the hour, Lloyd was performing a series of basic chemical reactions that bubbled, fizzed and smoked, transfixing his new friend. It was on this day, in Lloyd Jeffress' little Portland bedroom, that Pauling decided to become a chemist.

From that point on, the two boys were inseparable. When not at school or work, they were performing crude, and sometimes dangerous, experiments in the makeshift lab that Linus built in his mother's basement. Using donated or pilfered chemicals, the boys created noxious gases and exploding powders while dreaming of getting rich as corporate chemists.

Later, Jeffress played a key role in a crucial story that Pauling often retold: At the age of 15, Pauling had imagined himself as a chemical engineer, working for one of the United States' major companies. When Pauling told his grandmother this, Lloyd chimed in prophetically, saying, "No, he is going to be a university professor."

Following high school, Linus and Lloyd both attended Oregon Agricultural College, where Pauling studied chemistry and Lloyd majored in electrical engineering. Jeffress, however, developed an interest first in physics and later in the medical field, eventually graduating from the University of California with a PhD in psychology. Despite the divergence in their interests, the two stayed in intermittent contact for the following 60 years.

With Pauling at Caltech and Jeffress at the University of Texas in Austin, it was difficult for the men to meet with much frequency, but they visited one another as regularly as their schedules would allow, sometimes engaging in the hijinks of their youth. In a short manuscript written after Lloyd's death in 1986, Pauling recounted how they had deceived the guests at an academic event with Lloyd's "mind reading" abilities, a hoax successfully planned and orchestrated by the pair:

> "We were all in a quite large room in the psychology building, perhaps 35 feet by 60 feet. There were chairs all around the perimeter, and a long table in the middle. Over and over again Lloyd would go out of the room, Edward Tolman or some other professor would gather up a dozen objects from people in the room and put them on the table,

one of these objects would then be selected by someone, and everyone would be asked to look at the object at which this person was pointing and to remember which one it was. Lloyd was then asked to come back into the room. As he entered he would walk a few steps toward the table, then stop and stand still for a few seconds, then walk toward the table and around it looking at the objects, and finally would pick up the correct one, as his first choice. He repeated this action with a new selection of a dozen objects, perhaps half a dozen times, before giving up. The people at the party were, I think, getting more and more suspicious of me, but I do not think that they saw what was happening. Lloyd's scheme was a simple one. When he entered the room and stopped, counting the steps as a, b, c, ..., I was also counting the steps a, b, c, ..., and when he took the step corresponding to the initial letter of the selected object I, who had been sitting quietly before, would raise my hand to scratch my ear or would cross my legs or would move my body, as a sign to Lloyd that the proper initial letter had been reached."

In this same remembrance, he also tells readers of Lloyd's wedding, writing,

"In 1923 my wife and I were married in Salem, Oregon, the 17th of June. Lloyd served as my best man. [...] A few months later my wife and I were living in a small apartment in Pasadena when Lloyd and a young woman, a student in Berkeley, turned up, saying that they wanted to get married. I went to a church nearby, which I had not attended at any time, found the minister, and asked if he would come to our apartment and carry out the marriage ceremony. Accordingly, Lloyd and Sylvia were married in our small apartment, with my wife as Sylvia's attendant and me as Lloyd's best man. No one else was present except the minister."

Jeffress, like Pauling, was a highly successful academic. Though his career began slowly, the breadth and depth of his research expanded considerably as he aged, with the bulk of his papers produced after his 50th birthday. As an expert in experimental psychology, focusing on psychoacoustics, he served as the chairman of the University of Texas psychology department. His longstanding interest in physics also led to his teaching a few courses in the subject. Perhaps more surprisingly, his experience with wave transference resulted in work on mine-detecting devices for the United States military.

Over the course of his career, Jeffress earned a series of awards and commendations for his excellence as an educator and for his contributions to the field of psychoacoustics. Pauling personally took great pride in his friend's successes, expressing special interest in his scientific papers. "I have many friends," he concluded in his 1986 reflection, "but I continue to think of Lloyd Alexander Jeffress as my best friend."

9 Deer Flat Ranch

By Matt McConnell

Pauling harvesting abalone at Deer Flat Ranch, July 1963.

In 1955 Linus and Ava Helen Pauling headed to Berkeley, California from their home in Pasadena to attend a meeting of the Scientific Advisory Board of the Helen Hay Whitney Foundation. On the drive back

from this event, the couple decided to take the scenic route along Highway 1 down the California coast. Passing through the Big Sur area, Pauling noticed a point of land projecting into the ocean with a cabin and
barn and a herd of grazing cattle. He suggested to his wife that such a
location would be ideal as a country home for rest and relaxation. Ava
Helen smiled and directed his attention to a 'For Sale' sign on the side
of the road.

Middle-aged and famous, Pauling's busy lifestyle had fostered a
growing desire for a place to think without distraction. The Big Sur property, called Deer Flat Ranch, seemed the perfect location. A 163-acre
cattle ranch spanning a half-mile stretch of rugged coastline between
Soda Spring Creek and Salmon Creek — about twenty miles north of
San Simeon, and just north of Salmon Cone at Piedras Blancas — the
property was surrounded by National Forest land.

Captivated, the Paulings wanted to visit the ranch for a closer look,
but the owner of the land — a homesteader by the name of Walter Ray
Evans — was in the hospital and thus not available to arrange a personal
tour. However, Mrs. Evans granted the Paulings permission to return
to the property for an evening, and so in 1956 the pair drove back to
Salmon Creek and stayed the night, setting up camp near the barn and
sharing a sleeping bag underneath the stars. This visit must have made
a positive impression, because the Paulings purchased Deer Flat Ranch
not long after, in August 1956. Escrow documents that Pauling filed into
his personal safe indicate that the couple paid a total of $29,000 for the
property.

The ranch that the Paulings had purchased did not feature much
in the way of infrastructure, just a small cabin and a weather-beaten
barn for horses and equipment. Walter Ray Evans had built the cabin
in 1906 out of lumber that was floated in from offshore to a beach on
the property called China Camp. Six years after building the structure,
Evans moved the "Old Cabin" up the hill, so that the residence would
be nearer to the barn and also less susceptible to the pack rats and water
problems that had plagued the space at its beachside location.

Other than the barn, the Old Cabin remained the only habitable structure on the property until 1964. It was very small, consisting of just a single room, and housed a butane tank, a hot water heater, a miniature refrigerator, and a sparse assortment of well-used furniture. The nearest bathroom was located outside under a shaky lean-to. Electricity was usually available, but there was no phone service. After the Paulings purchased the property, they moved in an antique, wood-fired, cast iron stove that was forged in Oslo, Norway in 1825. This centerpiece of the humble home quickly became very popular with visitors.

While mostly a sanctuary from a heavily cluttered calendar, Deer Flat Ranch also represented an entirely different lifestyle in which the Paulings could explore new types of ventures. With the ranch came a herd of cattle, and within a year of buying the property, Pauling began pursuing an expansion and reorganization of his land in coordination with nearby private, state, and federal owners. Pauling's aim in doing so mostly revolved around his desire to extend the grazing area available to his animals. From the time that the ranch was purchased, Pauling paid $26 a year for grazing rights on adjacent Santa Lucia National Forest land, and also paid a neighbor named Patrick Boyd for additional grazing on his property.

In 1958 Pauling approached the local head ranger, Alexander Campbell, about the possibility of trading forestland to the north of Deer Flat Ranch for land northwest of the Salmon Creek Ranger Station. Specifically, Pauling wanted to trade 40 acres of his own property for 42 acres of forestland, the end result being a new northern boundary — Soda Spring Creek — for the ranch. These negotiations were conducted largely through Dale Haskin, who was the ranger working directly underneath Campbell at the nearby station.

Haskin had become close with the Paulings, at one point teaching Linus and Ava Helen's oldest son, Linus Jr., to wrangle, castrate and brand calves. By 1960 Pauling had hired Haskin as a ranch manager, a job that also involved supervising the property's itinerant worker, Phil Collum. A self-described colleague of author John Steinbeck — who himself was a native of nearby Salinas, California — Collum claimed to have traveled with Steinbeck up and down the West Coast during their younger years.

When the Paulings arrived at Deer Flat Ranch, Collum was found to be living on the property. Rather than evict him, the couple chose to furnish their newfound neighbor with a tent, and also offered a campsite that was suitably far away from the Old Cabin. Enabled by this offer of space and a $120 monthly paycheck, Collum continued to live at Deer Flat Ranch for many years, subsisting in part on local abalone (which he gathered from the beach) and red wine. He earned his monthly wages by working on the ranch, caring for cattle, making repairs, and cutting wood.

Although Linus and Ava Helen didn't often work directly with the cattle, Pauling acted as a head manager of sorts for the entire operation, keeping detailed employment records as well as notes on the current stock. Soon the Paulings were sending a portion of their cattle to market, while retaining others as a natural mechanism for mowing their grass. Each year, Linus Jr. and Ralph Haskin branded and castrated the new calves, with Collum and sometimes Pauling himself assisting with the wrangling. They then shipped the calves by truck to an auction house in Santa Rosa where, after they were purchased, area ranchers would fatten them up for market.

Pauling's experience of the cattle-ranching life had its moments of drama. In 1959 he noted in his research notebook that cattle rustlers were on the move in the area. Driving a white Ford sedan that was pulling a horse trailer into the mountains, the rustlers would shoot the cows with a tranquilizer gun, dress the meat, and pack it out. In 1961 a very arid summer ushered in soaring temperatures, and with it the grass and nearby fresh water sources dried up. That year only three steers were sold from Deer Flat Ranch, while 22 were found dead, including six young calves. Other "excitement" included a 1972 brush fire at the property.

In 1976 a neighboring rancher based in King City, California began grazing his cattle illegally at Salmon Creek. When the rancher "played dumb" in response to local investigations into the issue, Pauling contacted the offender directly and ordered him to personally fund and build a fence to keep his cattle contained. The strategy worked, perhaps due to Pauling's implied threat of a lawsuit.

The ranch also afforded other business opportunities for the Paulings. Most notably, in 1957 Pauling purchased an additional five acres at Piedras Blancas, about 20 miles south of the main property. The land was right on the beach, just off the highway, and came equipped with a small house and a gas service station. Linus and Ava Helen paid $14,000 for the parcel, which was purchased, once again, from Walter Ray Evans and his wife.

The service station was subsequently leased to Luther Williams, whom the Paulings also hired as a part-time ranch manager. Later on the station was rented out to a Mr. Mel Valois and his wife, who sold the gas to Chevron. By early 1958, Pauling was leasing the property to the couple for two cents per gallon of gasoline sold monthly, plus $338 in rent. The couple left the service station in 1962, but were quickly replaced by new tenants.

Between tending to the cattle at the ranch and operating the filling station, the Paulings continued to employ multiple part-time ranch managers and groundskeepers, with new employees cycling in and out every few years. Anywhere between three and five workers remained on the payroll until the late 1980s. By then, Ava Helen had passed away, Linus was well into old age, and the number of Pauling-branded cattle sold at the Templeton livestock market had dropped precipitously.

*

Stray incidents aside, Deer Flat Ranch quickly fulfilled its potential as a refuge from an extremely busy existence. A few years after buying the property, Ava Helen told her husband, "Do you know, we have been here for one week, you and I, without seeing a single other person? This is the first time in our 40-odd years of marriage that this has happened."

More than a refuge even, the ranch gradually emerged as a kind of paradise for the Paulings. One could reliably harvest ten abalone off the adjacent rocks at low tide, and Linus found that he greatly enjoyed collecting these sea snails with his wife, pounding them shoreside to make them more tender for dinner. In the pasture, a horse and a goat kept the cattle company, and marine life including otters and sea lions frequented the beaches. The Paulings also found satisfaction in collaborating on landscaping chores at the ranch, a pleasure that continued for Linus even after a 1960 incident that resulted in poison oak rashes on both arms.

During his solo trips to the property, Pauling frequently withdrew into the life of the mind. Pauling's literary and intellectual interests ranged far and wide, and his leisure reading included the poetry of the Greek atomist Lucretius, the rhetoric and philosophy of the great Roman orator Cicero, and the metaphysical proto-evolutionary poetry of Charles Darwin's uncle, Erasmus Darwin. Pauling's Deer Flat reading list also included a history of British chemistry, as well as Bertrand Russell's essay, "In Praise of Idleness," within which Pauling underlined the quote, "A busy man doesn't think."

Pauling also made note of re-reading Frederick Metcalf Thomas's *Estragia para la Supervivencia*, a work that had evolved out of Thomas's thesis. Pauling had read the thesis several years earlier and had even recommended it to Albert Einstein, who followed up on Pauling's tip and liked it so much that he subsequently wrote the preface for the book version of the text. While going through the work again at Deer Flat Ranch, Pauling underlined another quote that surely resonated with him: "The enslavement of scientists will not provide a solution for world problems."

Though Pauling clearly understood the importance of relaxation, work was never far from his mind, be it chemistry, medicine, or world affairs. By 1962 Pauling was writing the third edition of his successful textbook, *College Chemistry*, entirely at the ranch, typically devoting one week per month to the project while at the Old Cabin, undisturbed by the outside world.

In the 1960s and 1970s, Pauling also spent his time at the ranch thinking about a wide range of problems in chemistry. Among them were the promotion energy of hydrogen atoms; dihedral angles in H_2O_2

and other molecular structures; the stability of the N_2 molecule; electron bonds; antiferromagnetic theory; and much, much more. The bulk of Pauling's research notebooks from this period consist of musings on current papers in chemistry representing significant problems; he seemed intent on deducing the solutions to all of them, sitting in his cabin with nothing but a pen, paper, slide rule, and the crashing of the nearby waves.

*

From 1956 to 1964, the Paulings stayed in the Old Cabin when at the ranch, and for much of this time, visiting family members would often sleep in the barn. By 1961 a pre-designed kit home had been constructed for guests to use. Located down the gorge from the barn, at the foot of Salmon Cone, the house came to be called China Camp, named after the adjacent beach.

That same year, Pauling began conversations with his former student, Dr. Gustav Albrecht of Caltech, about acting as chief architect on the design of a new home on the property. Albrecht worked with John Gamble Associates and, once construction began, lived in the Old Cabin for several months to supervise the building according to Pauling's specifications.

By 1964 the "Big House" was complete. It was an unorthodox space, filled with angled windows of multiple types that offered numerous views of the Pacific Ocean. The home likewise featured dueling his and hers studies, as well as a garage that was specifically built to shelter a car while also housing Pauling's collection of scientific journals. Bookcases were everywhere and, inevitably, the decor came to be dominated by framed honorary doctorates lining the hallways and mounted in every room. A massive stone fireplace separated the kitchen from the living room, and a large westward-facing deck became a focal point for social gatherings. The unusual residence proved difficult to maintain, but for the Paulings it was heaven on Earth.

The Big House was built at the end of a new road that had been bulldozed west from the Old Cabin to a nearby glade that the Paulings called "Eucalyptus Hollow." By the time that most of the construction was completed, excavations on the Pauling land had revealed that a small village of Salinan native people, dating back thousands of years, had once been located in the area where the Big House was built. More artifacts were discovered under the Old Cabin, which was rebuilt and, a decade later, joined by a new caretaker's house.

Kids and grandkids visited during and immediately following the construction, and were often enlisted to perform various duties on the property. Linus Pauling Jr., his wife Stephanie, and Stephanie's daughter Carrie were frequent visitors throughout the 1970s. During this time they assisted with finishing floor moldings and tiles at the Big House, which sported a decorative copper diving screen based on the mezzanine foyer in Dulles National Airport, as well as a specially made copper roof.

The Big House was a sanctuary, and it was understood that even family visitors were not to barge in unannounced. Rather, Ava Helen would run a dish towel up a nearby flagpole when she was ready to receive visitors, usually in the late morning. During visits with family, Pauling tended to focus his conversation on scientific matters, while it was Ava Helen who worked to bring the family together, particularly relishing her role as a grandmother. Catching fish from the Pacific Ocean and cooking under gas lights in the wilderness of California's coastal forests, visitors often felt a sense of living in the pioneer past. Linus and Ava Helen reveled in this sensation themselves, and after 1966 they were spending fully half of their time at the Big House.

With advancing age came thoughts of retirement, and Linus and Ava Helen began to imagine that they might move to the ranch full-time by Pauling's 70th birthday, or perhaps his 75th. Pauling was, unsurprisingly, consistently non-committal about the idea of giving up his career in science, no matter how old he grew. In the meantime, the ranch offered an appealing happy medium where he could continue to pursue a scientific agenda while lessening the pace and clutter of his very public life.

By 1976, when Ava Helen was diagnosed with cancer, the pair seemed to treasure their time at Deer Flat Ranch all the more; Ava

Helen practiced the guitar and bought a grand piano, and family came to visit often. Pauling, however, could never truly be removed from science, and spent much of his time at the ranch working on theoretical papers. A year later, Pauling looked to be making good on notions of retirement, and was considering removing himself from day-to-day operations at the Linus Pauling Institute of Science and Medicine (LPISM), which Art Robinson was leading as president. However, an administrative battle with Robinson that arose over the future of the Institute provided Pauling with a compelling reason to remain heavily involved (see Chapter 20), and he never did fully extricate himself from LPISM affairs.

In retrospect it seems unlikely that a man of Pauling's industry, interests and ego could totally remove himself from the world of science and retire to a ranch full-time. In fact, after Ava Helen passed away in 1981, he ramped up his scientific program, working mostly at the ranch and in Palo Alto, California for the remainder of his life. Alone and growing older, the rustic pioneer charm of the ranch began to fade somewhat for Pauling. The land around the gas service station was found to be eroding at an alarming rate, and it was eventually abandoned. Likewise, the number of cattle and ranch hands slowly dwindled. Eventually, only a single caretaker remained on the property: Steve Rawlings, who also acted as Pauling's personal nurse during his final years.

Nonetheless, Pauling continued to spend a majority of his time at the property, reading, writing, and dreaming of a peaceful world guided by the light of scientific reason. It seems fitting then that when Linus Pauling passed away, in August 1994, it was in the Big House at Deer Flat Ranch, surrounded by his family.

10 Stuck on a Cliff

Chapter

By Megan Sykes

Pauling Saved From Coast Cliff

Associated Press Wirephoto
Dr. Linus Pauling, Nobel Prize winner, drinks hot coffee
after being found on mountainside near Monterey, Calif.

On the morning of January 30, 1960, a Saturday, Linus Pauling told Ava
Helen that he would be out checking the fence lines along the boundaries
of their ranch near Big Sur, California. A little before 10:00 A.M., Ava

Helen watched as her husband walked towards the coast south of their cabin, but did not notice — as Pauling mistakenly believed she had — as he veered away from the fence line and toward Salmon Cone, a small mountain in the Santa Lucia chain near the Pauling home. Linus was dressed comfortably, wearing slacks, a light jacket, and his signature beret. He and Ava Helen had planned to meet for lunch and thought that a friend would perhaps stop by around noon; all expected Pauling to be back by that time.

For several years, Pauling had been interested in finding the mouth of nearby Salmon Creek. Thinking that it might be located around the China Camp area of the property, he climbed several ledges that allowed him to walk east to investigate his theory. After a time he found and followed a deer trail, rising several hundred more feet in elevation. When the trail came to an end, Pauling thought that he saw a spot 20 or 30 feet above him where the path picked up again, and he began trying to make his way to it over loose rocks. Quickly finding that he was not in a position to move east as he had hoped, Pauling saw that he was both unable to retrace his steps back down or to go further up the slope safely. A sickening realization washed over him: he was stuck.

The ledge on which he was perched measured about three feet by six feet. Loose rocks, leaves, and sticks covered the spot and behind him was a sheer rock face. Pauling sat on the ledge for several hours thinking that Ava Helen would walk along the beach and see him stranded. But as afternoon moved into evening, he came to realize that he might have to stay the entire night perched on the little cliff. Having reached this conclusion, he began to dig a little hole with his walking stick in which he could sit. This hole took the form of a two foot by three foot rectangle that was about a foot deep. He then used the extracted dirt to create an eighteen-inch mound around the hole. His resting area completed, Pauling intently pondered a route off the ledge, only to become too frightened to continue on his own. Soon it was dark.

Pauling did not want to fall asleep during his long night on the cliff because, for one, he was afraid he would not hear the calls of searchers.

More importantly, Pauling was very concerned that, in the midst of sleep, he might roll off the precipice and into the crashing ocean below. In order to remain awake, he engaged in a variety of mental tasks. For a while, he lectured to the waves about the nature of the chemical bond. He also listed the various properties of the elements of the periodic table. As the night dragged on, Pauling counted as high as he was able in as many languages as he could — German, Italian, French and eventually English. He even used his walking stick to try to tell time based on the positions of the constellations. In an effort to retain heat and stay alert, Pauling tried to move one limb or another at all times.

Moving his arms and legs was only part of Pauling's strategy for keeping warm on this January night. Earlier in the evening, having decided that it would be necessary and prudent to remain on the cliff in the small hole that he had dug, Pauling began pulling up some of the bushes that were located near his little ledge. He broke them into smaller pieces and placed them on the damp bottom of the hole. He then laid some of the intact branches over himself. This tactic proved unsatisfactory however, as he had to constantly pull out small leaves and twigs from inside his clothes; eventually he just broke the bushes into twigs, which he used as both mattress and blanket. Wishing to be warmer still, Pauling unfolded the map that he had brought with him and laid it over himself. He later told his family that the map had helped immensely. Luckily, it was also not as cold a winter night as could have been the case.

Well before Linus began settling in, an alarmed Ava Helen had sprung into action to find him. When he missed their lunch date, Ava Helen had assumed that he had simply lost track of time and did not worry too much. But when 4:15 P.M. rolled around and there was still no sign of her husband, she went to the nearby ranger station to ask for help and to call her son-in-law Barclay Kamb. (The Pauling home, at that time, was not equipped with a phone.) As it happened, a ranger had come close to Pauling's ledge near Salmon Cone, but the trapped scientist was unable to attract his attention and the ranger moved on to search other areas. At 11:30 P.M., a deputy sheriff from Monterey called off the search for the night as the weather conditions were not conducive — intermittent clouds and fog enshrouded the area from early evening until well after Pauling had been rescued.

Undeterred, Barclay Kamb reached the ranch around 2:30 A.M. and began searching for an hour and a half in the direction that Ava Helen had last seen Pauling...the wrong direction. At 6:00 A.M., he resumed looking in the same area again. Just before 10:00 A.M., Pauling finally heard another of the searchers; a man named Terry Currence, who was walking along the beach below the ledge. Pauling called to him and Currence scrambled in Pauling's direction. Currence then called to the deputy sheriff, who maneuvered to a spot a little ways above Pauling. Currence was sent by the deputy sheriff to fetch some rope and to tell Ava Helen that her husband had been found alive and well. While waiting for the rope, Pauling and the sheriff successfully made their way off the cliff, using one of the paths that Pauling had been afraid to follow unassisted.

Pauling was in good spirits as he was led back to his residence, even joking with the rescue team, and upon returning home he had lunch and some coffee. Meanwhile, Ava Helen shooed away the reporters who had assembled and thanked everyone who had helped to find her husband. The following day, the couple packed up their car and drove back to Pasadena.

On Tuesday, Pauling went to Caltech to teach a class. When he finished lecturing, he walked past the small party that his colleagues had put together to welcome his return, and went into his office without saying a word. He locked his office door and shoved a note underneath it requesting that his schedule be cleared for the day. His staff was unsure of what to do so they called in Barclay, who came to Pauling's office and drove him home.

Once he arrived home, Ava Helen put her husband to bed and called the doctor. Pauling was diagnosed as having gone into mild shock and was told to rest for several days. He was likewise afflicted with a severe case of poison oak, an unfortunate side effect of his makeshift bedding on the ledge. Pauling remained in bed for several days and barely spoke, crying at the sight of his grandchildren when his daughter Linda brought them over for a visit. The trauma was relatively short-lived though, and two weeks later he was not only talking and responding to letters but also honoring speaking engagements again.

The media response to Pauling's plight on the cliff had been swift and extensive. By 9:30 A.M. on Sunday, news of Pauling's disappearance had spread across the radio, and a half hour later, at 10:00 A.M., an overzealous reporter told San Francisco Bay Area residents that Linus Pauling was dead. Two of Pauling's children, Linda and Crellin, were informed of the radio broadcast, and for an hour were unable to discern otherwise — they thought their father had died in a freak accident.

After Pauling was found, news reports of the past weekend's events spread around the world, from Oregon to Massachusetts, India to Australia. Over the coming month, Pauling received a stream of well wishes from colleagues, friends, family, and even strangers who had heard of his ordeal. One such telegram read as follows: "Dear Dr. Pauling, Will you be so kind as to stay off precipitous cliffs until the question of disarmament and atomic testing is finished? A needy citizen. [Signed] Marlon Brando."

11 The SISS Hearings

By Will Clark

PAULING REFUSES — Dr. Linus Pauling, Nobel prize winning physicist, refuses to give a Senate internal security subcommittee the names of scientists who helped him circulate petitions asking nuclear arms ban. Pauling did disclose the signers.
(P) Wirephoto

Pauling Again Refuses Names to Senate Quiz

As was the standard for him after World War II, Linus Pauling's time in the 1950s was divided, frenetically, among research, public speaking and social demonstration. As the decade progressed, he remained actively engaged in academia, but was directing more and more attention to nuclear non-proliferation issues. One of Pauling's main concerns as an activist was the tepid response from public officials and the Atomic Energy Commission to what he viewed to be a major problem: radioactive fallout from atomic bomb tests. After conferring with two scientific colleagues, Barry Commoner and Edward Condon, it was concluded that, because of their insight and technical knowledge of the dangers involved, the nation's scientists bore a special responsibility to speak out about nuclear testing.

As a result, Pauling and several associates began circulating a petition. Copies were distributed to individual scientists across several states, and soon a large collection of researchers from numerous universities and institutions began responding en masse. The petition's purpose was made very clear in its introduction: "We, the American scientists whose names are signed below, urge that an international agreement to stop the testing of nuclear bombs be made now."

The initial document, "Appeal by American Scientists to the Government and Peoples of the World," came back to Pauling adorned with over 2,000 signatures, including those of several prominent members of the scientific community. After a short interlude, the petition was supplemented by an international version, ultimately raising the tally to more than 11,000 signatures by the time of its submission to the United Nations in early 1958.

The endeavor was seen as a huge success by advocates, but it also instigated a new movement against Pauling, one propelled by several public agencies and officials. In particular, the Senate Internal Security Subcommittee (SISS) began focusing more attention on Pauling, though it was not the first time he had found a place on their agenda. In 1955 the SISS released a tract titled "The Communist Party of the United States of America: What It Is, and How It Works." Linus Pauling's name was on a list of individuals said to be among the most active participants and supporters of Communist fronts.

Similarly, when Pauling and his associates released the first version of their nuclear test ban petition in June 1957, they were met with substantial criticism from a wide spectrum of opinion-makers. Initially Pauling's scientific authority on the issue was the primary point of contention. Soon enough though, Pauling was being accused of Communist conspiracy, and was subpoenaed by the SISS to discuss the potential role of Communist organizations in the petition's distribution. Pauling expressed his willingness to appear before the subcommittee, but unforeseen senatorial politics eventually interceded, forcing a temporary delay of his compulsory appearance in Washington, D.C.

The SISS itself was essentially the Senate version of the House Un-American Activities Committee (HUAC), and had served similar purposes in the past. Indeed, in 1960 the committee was composed of many staff members recycled from past HUAC activities. Most notably, Senator Thomas Dodd, chairman of the subcommittee at the time, conferred regularly with former HUAC investigator Benjamin Mandel. Senator Dodd and others were also working periodically with the Federal Bureau of Investigation (FBI) in a joint effort against suspected Communist subversion.

Before sharpening their focus on Pauling, Dodd and the SISS were popularly credited with instigating the dissolution of the National Committee for a Sane Nuclear Policy (SANE), the umbrella organization for a collection of locally oriented anti-bomb protest groups. Dodd worked discretely with SANE's national leadership in providing the organization with an ultimatum: either cull itself of alleged Communist membership or risk a prolonged investigation by federal authorities. SANE's leadership chose to require loyalty oaths, a move that split the collective, leaving the severed parts largely incapacitated. After the subcommittee's apparent victory, Dodd and his counsel were emboldened enough to focus their full attentions on Pauling.

Following a speech given to the Women's International League for Peace and Freedom in the spring of 1960, Pauling was handed several fliers, a poem, some news clippings and other random papers while answering people's questions. That night, after returning to his hotel room, he found a subpoena addressed to him within the clutter. The order adjured his appearance before the SISS on June 21st, two days from then, to address: "Communist participation in, or support of, propaganda campaign against nuclear testing, and other Communist or Communist-front activity with respect to which you may have knowledge."

Abraham Lincoln Wirin, a lawyer who had helped Pauling through an earlier dispute, flew directly to Washington, D.C. the next day. The duo discussed Pauling's options and decided to utilize the press as much as possible, a tactic that had proven fruitful for Pauling in the past. Though Pauling was given very short notice, he and Wirin were able to devise what seemed to be a simple but promising strategy: make the entire affair as public as possible, cooperate to whatever extent was appropriate, and maintain the integrity of Pauling's constitutional rights. Though he was alarmed, Pauling felt reasonably prepared for the engagement ahead of him.

*

"I gladly confess that I recall Tom Dodd as a somehow larger and more appealing figure than his critics acknowledge or the record of the recent past shows. For twelve years I followed him, long enough to leave with me some fond memories and a share in his guilt. Had he been only an empty or venal man, his story would be unimportant. It is what he might have become, had the system through which he rose encouraged his strengths instead of his weaknesses, that gives to his fall an element of tragedy and hence a claim to significance."

— James Boyd, former staff member and associate

Pauling's perception of SISS chairman Dodd was strongly (and understandably) shaped by the unjust position into which he had been placed by the subcommittee. However, the true character and life of Thomas Dodd, as revealed by those who knew him personally, is full of complexity and provides a useful illustration of the practices prevailing among U.S. government officials around the middle of the previous century.

Thomas Joseph Dodd was born in Norwich, Connecticut on May 15, 1907. Growing up in New England, he attended public schools and was raised in the Roman Catholic Church throughout most of his childhood. He earned his bachelor's degree from Providence College, and received a law degree from Yale University in 1933. He married a year later, and helped raise six children.

Dodd worked at the FBI as a special agent for one year, before being appointed Director of the National Youth Administration for Connecticut. He later served as Special Assistant to the Attorney General, during which time he was best known for his role in the prosecution of Ku Klux Klan members in South Carolina and for defending labor rights in Georgia.

During World War II, Dodd primarily worked on cases involving espionage and sabotage, but also played a role in uncovering fraud committed by American industrial firms. Following the end of the war, he was solicited to aid in the prosecution of Nazi war criminals at the Nuremburg trials in Germany. Active in virtually all aspects of the prosecution, Dodd served as vice-chairman of the Review Board and Executive Trial Counsel, essentially making him the second-ranking lawyer and supervisor for the U.S. prosecution team.

After returning from Europe, Dodd entered into private practice and was subsequently elected to the U.S. House of Representatives in 1952, after an unsuccessful run for governor of Connecticut four years prior. In 1958 he was elected to the U.S. Senate, serving on the Judiciary Committee and the Senate Foreign Relations Committee. As a component of these duties he co-chaired the SISS, the junction at which he ultimately crossed paths with Linus Pauling.

Dodd was known by colleagues for his low level of partisanship, but he was more widely recognized for his stance as a strident anti-Communist. (The journalist Drew Pearson referred to him as a "bargain-basement McCarthy.") His worldview was fundamentally inspired by notions of loyalty, and he made a name for himself as one who remained faithful to beleaguered political figures in times of crisis. To his family and those close to him, he was admired for his mirth, humor, generosity and capacity for fellowship, and during his career as a public servant, he often felt he was upholding a strict moral code. Despite his seemingly lofty ideals and persona however, Dodd faltered as he became increasingly intermingled in political graft.

Though the events that transpired in 1960 with Linus Pauling proved a major event, the time he spent on the SISS was seen by supporters and critics alike as a comparatively minor chapter in his life. He is instead more widely recognized for his notable rise through U.S. institutions, and for his misconduct with regard to government ethics violations.

After violating flaccid campaign finance laws and abusing government funds, Senator Dodd was involved in a drawn-out congressional inquiry that concluded with his censure by the U.S. Senate. Prior to the vote by his peers, a clerk read out the Censure Resolution, which included the following: "It is the judgment of the Senate that the Senator from Connecticut, Thomas J. Dodd...deserves the censure of the Senate and he is so censured for his conduct, which is contrary to accepted morals, derogates from the public trust expected of a Senator, and tends to bring the Senate into dishonor and disrepute."

Senator Dodd's rebuke, which was issued seven years after his confrontation with Pauling, made him the first Senator in U.S. history to be censured for financial misconduct. (He was also the first Senator to be censured since Joseph McCarthy in 1954, and one of only six Senators censured over the whole of the 20th century.) Though the possibility of criminal investigation from the IRS and Justice Department loomed overhead, Dodd carried on as though nothing had happened. Ignoring an outcry for his resignation, he remained on his committees and continued to exercise his basic senatorial duties.

Ultimately, the Senate chose not to pursue any further investigation, instead concluding their formal report with a reprimand aimed at Dodd's accusers, many of them former staff members. Dodd was allowed to complete his term largely undisturbed, wielding the waning power that remained within his grasp. The censure vote did have lasting effects for the system as a whole however, by publicly demonstrating the need for legislative ethics law reform. Reflecting on the time period, a former aide offered an explanation for Dodd's fraudulent conduct, and the Senate's hesitant enforcement:

"The Senate appeared to Dodd not as a harsh and exacting judge, but as a permissive and protective accomplice. His occasional inanities in debate would always appear in the Record as words of wisdom. His absence would be reported as a presence. His vacation trips would not only be paid for by the Senate, but would be billed as 'official business.' His honorariums and legal kickbacks and finder fees and gifts were excused from the prohibitions that covered all government officials except Congressmen and Senators. His fraudulent campaign reports would always be accepted at face value. Not only could he keep unsavory contributions, he could route them through the Senate Campaign Committee and thus hide their origin. He could use his official allowances to buy birthday presents, wedding invitations, and the like, and no one would know. He had reason to consider himself immune from investigation."

Spurned by the Democratic Party, Dodd lost in his bid to run as an Independent in the 1970 Senate re-election campaign. Less than one year later, on May 24, 1971, he fell victim to a heart attack, dying in his home at the age of 64.

*

"It is well, therefore, to begin with the simple legal profile of what a Committee can and cannot do. All that it can do is to compel a witness to testify. If he talks, and talks honestly, that is the end of the matter. The Committee has no power to deal with him further; it is not adjudicating or determining any of his rights or duties; it is simply collecting information for legislative use. If a witness does not answer honestly, he may expose himself to a perjury prosecution; if he refuses to answer at all, he may expose himself to a contempt prosecution."

— Harry Kalven Jr., 1960

Pauling was ordered to appear before the SISS on June 20, 1960. He and his counsel arrived on time, as requested, at the New Senate Office Building in Washington, D.C. but the Senate was in session, and Pauling's hearing was postponed until the following morning. Already inconvenienced by the delay, Pauling was likewise, that day, notified for the first time that the hearing would be an executive session, and thus not open to the public or the press. Importantly, this meant that Pauling would not receive a copy of his testimony, and that the subcommittee reserved the right to release portions of Pauling's testimony to the public at their own discretion. The subcommittee defended the decision by suggesting that "It is consistent practice...for the protection of its witnesses to hold executive sessions in advance of public hearing for the purpose of siting evidence, checking its accuracy and elimination of non-evidential material."

Though the delay further aggravated his disposition, Pauling made good use of the extra time, appearing before the press and giving several interviews. That evening he also sent a telegram to Senator James O. Eastland, chairman of the Judiciary Committee, of which the SISS was a subcommittee. In it, Pauling made clear that the subcommittee's stated concerns for keeping the hearing closed — ostensibly to protect his rights as a witness — were unwarranted, writing, "When I appeared this morning as commanded I was surprised to learn for the first time that the session, postponed by you until 8 A.M. tomorrow, is to be an executive hearing closed to the public and the press. I do not like secrecy and I wish to present my testimony in public. I have nothing to say that needs to be kept secret and I neither require nor desire the protection of an executive hearing."

His protests were treated sympathetically in several prominent newspapers the following morning and, on June 21, after a few minutes of inquiry in executive session, Senator Dodd reversed the earlier decision and opened the hearing to the public. The first moments of the open session were nonetheless used by Dodd to once again justify his initial decision to hold an executive session. After defending himself satisfactorily, and explaining the rationale for Pauling's presence before the subcommittee, Dodd made a statement that clarified his fundamental intentions:

> "...while the particular objective of the session today is to learn what we can from this witness respecting Communist activity in connection with protests against nuclear testing, we shall also seek other information respecting Communist activity if it appears that such information might be available from this witness."

Having established the subcommittee's privilege to address the matter, Dodd briefly placated Pauling before moving on with his agenda. From there, Chief Counsel Jules Sourwine took control of the proceedings; throughout, it was Sourwine who would conduct the majority of the questioning, allowing Senator Dodd to play a peripheral but still authoritative role.

Sourwine soon made it clear that the main topic of interest for the hearing was the nuclear test ban petition that Pauling had submitted to the United Nations more than two years earlier. After reading aloud the press release that had preceded Pauling's petition submission, Sourwine began asking about the signature collection process that Pauling had utilized in compiling the petition. The subcommittee had previously acquired a copy of the document from the United Nations, and after determining that Pauling had not received each signer's name on separate sheets of paper, Sourwine requested from Pauling the names of those individuals who had collected and turned in more than one signature.

After repeatedly avoiding a direct answer to the request, Pauling was finally forced to fully face the issue. Having cooperated with little reservation up to this point, Pauling politely but firmly refused the request, defending his decision as a matter of conscience and stating that, "I feel some concern about my duty to the people who worked for this petition. I feel concern that they may be subpoenaed before this Subcommittee, subjected to the treatment that I have been subjected to."

Senator Dodd justified the request by mentioning discrepancies that his staff had encountered with names that Pauling claimed were on the petition, as well as the number of signatures reported by a United Nations staff member. Following the hearing, it was determined that responsibility for the error in reporting the total number of petition signatures resided with the UN, but the allegations during the hearing still served to call into doubt Pauling's penchant for telling the truth.

The subcommittee then addressed the formation of the petition, the sources of its funding, and the nature of the roles played by Russian and other international scientists. From there, Chief Counsel Sourwine spent the final phases of the morning session inquiring into Pauling's affiliations with suspected Communists and Communist fronts, including the National Council of the Arts, Sciences, and Professions, and the organization that had printed the blank petition forms.

Finally, as noon approached, Sourwine again requested both a list of individuals that the petition had been sent to and a list of the individuals who had delivered multiple signatures, with an indication of how many signatures were sent in. The hearing was recessed, and Pauling was given a chance to consult with his counsel, A.L. Wirin, before responding to the subcommittee's final requests.

*

When the hearing resumed two and a half hours later, Pauling agreed to submit a list of the individuals that he had sent petition requests to, but refused to submit a list of those who returned more than one signature. In making this stand before the Senators, he spoke eloquently of his decision to refuse the subcommittee's request:

"The circulating of petitions is an important part of our democratic process. If it were to be abolished or greatly inhibited, our nation would have made a step toward deterioration — perhaps toward a state dictatorship, a police state.

"I am very much interested in our nation, in the United States of America, and in the procedures that were set up in the Constitution and the Bill of Rights. Now, no matter what assurances this subcommittee might give me about the use of the names of the people who circulated the petition that I wrote, I am convinced that these names would be used for reprisals against these believers in the democratic process; these enthusiastic, idealistic, high-minded workers for peace. I am convinced of this because I myself have experienced the period of McCarthyism and to some extent have suffered from it, in ways that I shall not mention. I am convinced of it because I have observed the workings of the committee on Un-American Activities of the House of Representatives and of this Subcommittee on Internal Security of the Judiciary Committee of the Senate. I feel that if these names were to be given to this subcommittee the hope for peace in the world would be dealt a severe blow. Our nation is in great danger now, greater danger than ever before... This danger, the danger of destruction in a nuclear war, would become even greater than it is now if the work for peace in the world, peace and international law and international agreements, were hampered.

"A terrible attack is being made now in the United States on the efforts of our government to achieve international agreements for stopping the bomb tests and for disarmament. This attack is being made by representatives of defense industries who benefit financially from the Cold War... I believe that the work for peace and morality and justice in the world needs to be intensified now, and I plan to do whatever I can in working for peace in the world, working for international agreements about disarmament."

Though his counsel supported his position on constitutional grounds, Pauling justified his defiance "as a matter of conscience, as a matter of morality, as a matter of justice."

Upon hearing his decision, and after giving Pauling one more chance to comply, Senator Dodd issued a formal statement of disapproval and dissent against Pauling's objection. After other senators on the dais had been given time to voice similar displeasures, Pauling was ordered to appear again at the New Senate Office Building in less than two months' time. Once more, he was directed by Senator Dodd to bring "all signatures or purported signatures to the petition...together with all letters of transmittal by which, or in connection with which, such signatures were transported to you or received by you." Implicit in Dodd's demand was the threat of a contempt of Congress charge, an offense that could lead to imprisonment should Pauling fail to comply. Once the order had been issued, the hearing was recessed.

Pauling spoke with the press after finishing his testimony, criticizing the implications of the subcommittee's requests, and justifying his refusal to turn over the list of signature collectors. Though Pauling had, by and large, met the committee with patience and cooperation, there were a number of issues that had yet to be addressed or accounted for, including several contentious sections of testimony that Pauling wished to read over. The tone of the hearing had turned somewhat sardonic by its end, and Pauling's indignant interpretation of the proceedings quickly made its way into headlines across the nation.

*

Shortly after the conclusion of his first Senate hearing, Pauling's counsel succeeded in postponing the scheduled follow-up from August to October 1960. The extra time gave Pauling room to plan for his upcoming defense, and to follow through on other plans he had made before being served with the SISS subpoena. In due course, he attended rallies and events to which he had previously been committed, making sure to discuss his dispute with Senator Dodd whenever a chance presented itself. He also encouraged people to write to their representatives in protest of his treatment, and gave many interviews critical of the proceedings.

Quickly it became clear that Pauling was gaining popular support, evident in part from positive mentions in editorials and letters to the editor. The *Washington Post*, for one, offered that, "Justice is best served at times by those who defy authority. Prof. Linus Pauling offered a splendid illustration of the point, we think, when he refused the other day to give the Senate Internal Security Subcommittee the names of persons who had helped him circulate a petition in favor of the abandonment of nuclear weapons tests."

Pauling felt that he had been subjected to a great injustice. Though he had remained calm and civil during the questioning, his anger quickly surfaced after the encounter. He resented the hearing in a general sense, and was particularly embittered by a few particular aspects of the experience. Once they were available, he took his time reading over the hearing transcripts, slowly digesting the implications of the subcommittee's line of questioning. After combing through his testimony, Pauling wrote a letter to the SISS that was eventually attached to his testimony, and thus made part of the public record. The letter expressed his sense of victimization, and detailed six specific points from the subcommittee's questioning that he found to be exceptionally inappropriate.

Pauling took special offense to the title that was given to his hearing, "Communist Infiltration and Use of Pressure Groups," a decision by the subcommittee that seemed particularly baleful. Pauling emphasized that the subcommittee had both no substantive reason to suspect that he was involved with Communism or Communist conspiracy, and that no new evidence connecting him to Communist activity had been revealed by the end of the hearing.

Another criticism that Pauling listed was a question posed by the subcommittee, which suggested that he had omitted Soviet signatures from the United Nations Bomb Test Petition when it was released to the press. Three months prior to the hearing, the subcommittee had in fact obtained a complete copy of the petition, which included the Soviet signatures. Pauling considered these three months ample time to clarify such an obvious error well before his questioning, and made his perception of the matter explicit, writing, "Damage was done to me by your false statement and by my having been questioned on the basis of your false statement. No matter whether the false statement was made (by Chief Counsel Sourwine) through gross carelessness or through malignancy, I protest this action."

A similar point of contention involved a letter that the subcommittee introduced into the record, which it had received from a staff member of the UN. It stated that, while Pauling had listed a certain number of petition signatures in his original press release, the United Nations had only received a portion of that number — an error that, as it turned out, was attributable to the United Nations. Though the staff member's mistake was eventually clarified for the record, Pauling was later informed that the subcommittee had been previously aware of the error. Indeed, a complete and accurate list of the names, matching those specified in Pauling's press release, had been in front of the subcommittee during the hearing itself. Pauling accused the subcommittee members of intentionally entering the untrue statement into the record in order to damage him, by defaming his reputation and casting doubt upon his integrity.

Lastly, Pauling addressed the subcommittee's threat, by intimation, of imprisonment. During the hearing, a subcommittee member had asked Pauling if he was familiar or acquainted with Dr. Willard Uphaus. The question was posed after Pauling had displayed a continued reluctance to reveal the list of individuals who had delivered more than one signature to the petition. Pauling answered that he did know of Uphaus, though he made no open recognition of the question's implication. He was surely aware however, that Uphaus was, at the time of the hearing, in jail for a transgression — contempt of court — similar to the one that Pauling was about to commit. Pauling closed his letter to the SISS with a biting critique of the question and its inference:

> "I consider this veiled threat, this intimation of the fate that awaited me if I did not conform to the demands of the Subcommittee, to be

unworthy of the Senate of the United States of America. My respect for Senator [Norris] Cotton would be greater than it now is if he had said straightforwardly that for me to refuse to give the Subcommittee the information demanded by it might lead to my citation for contempt of the Senate and to a prison sentence. I prefer straightforward statements of fact to veiled threats and attempted intimidation. I prefer the forthright search for the truth to the sort of trickery and misrepresentation that in my opinion has been revealed by the proceedings in my hearing before your Subcommittee."

On top of his continued interactions with the press and his efforts to revise the published testimony for the public record, Pauling also took direct legal action after the first hearing. In particular, he sought a declaratory court judgment that would affirm his right to refuse the subcommittee's request. In so doing, he was attempting to clarify his position and reduce his risk in the matter, but the District Court and Court of Appeals for the District of Columbia both ruled against him. He appealed and, at the start of his second hearing in October, his case was pending before the U.S. Supreme Court.

Pauling also continued to travel during the interlude between his hearings, visiting London and Geneva, where he furthered the discussion of an atomic test ban treaty with American, British and Soviet officials. He received a great deal of support during the trip, and the pressure from the hearing, as well as the threat of imprisonment, seemed to lessen as a result.

Though Senator Dodd attempted to address Pauling's growing campaign against the subcommittee, increasing his rate of public retaliation as the second hearing date grew closer, he found that he could not match Pauling's intensity and drive. Recognizing the seriousness of the situation, Pauling was trying very hard to avoid jail time. At the same time, he was defending his reputation (and thus his very livelihood) as a scientist, academic and activist. Dodd's political gambit for an increased share of national spotlight, however much buttressed by honest concerns about Communist subversion, was no match for Pauling's grasp of the situation's severity. As the second hearing drew closer, Pauling's resolve continued to strengthen. Pauling's supporters, on the other hand, remained anxious, as the outcome of his fate, which lay at the discretion of Senator Dodd and the SISS, remained elusive.

*

Pauling departed for Europe with Ava Helen shortly after his first engagement with the SISS. Though he was forced to shorten his trip because of complications associated with the upcoming second hearing, he still managed to spend three weeks in the U.K. and Switzerland. The trip overseas provided a mix of recreation, professional visits and diplomatic endeavor, but a primary motivation for the journey was to attend the Tercentenary Celebration of the Royal Society. Pauling flew to London on July 14, after which he attended a number of lectures and receptions, visited Oxford and went to the Glyndebourne Opera in Sussex. He also gave several talks about his most recent experiences, but spent most of his time discussing international disarmament and world peace.

While in Europe, Pauling visited with ambassadors from Great Britain, the Soviet Union and the United States, including a 40-minute appointment with the U.S. Ambassador to the United Nations, James J. Wadsworth. According to notes penned after the meeting, the ambassador was very supportive of Pauling's work and with his unwillingness to turn over the names of people who had circulated petitions. Wadsworth also shared the opinion that the viewpoint framed in Pauling's petition, when it was written three years earlier, was now official government policy. Wadsworth encouraged Pauling to continue working on the grassroots level, and also reassured him that no one in Washington could stand up against public opinion, a validation of Pauling's pre-hearing tactics to garner public support. The meetings, and the trip in general, seem to have lent Pauling comfort as he made his way back to Pasadena and the upcoming confrontation with Senator Dodd.

Meanwhile, press coverage following Pauling's first testimony remained largely sympathetic. In the months between the June and October hearings, many more articles were published which mentioned the case generally and provided a neutral presentation of the facts. A majority of the print media content, however, came in the form of letters to the editor, editorials, and opinion articles written from all over the country. While less prevalent, coverage of Pauling's trials also appeared in the international press, primarily in Europe.

The opinion articles written about the matter tended to take one of two positions — either framing Pauling as the victim of unscrupulous political prejudices or, less commonly, as a supporter and contributor to Communist causes. Norman Thomas, writing in the *Post War World Council Newsletter*, exemplified the first approach:

> "Dr. Linus C. Pauling has been subpoenaed to testify 'on Communist participation in, or support of, a propaganda campaign against nuclear testing.' The subpoena was issued by the Senate Internal Security Subcommittee. This gratuitous intrusion of a Senate committee is not born of any reasonable fear of Communist propaganda but rather of desire to stop any criticism of any resumption of nuclear tests. The committee might more usefully subpoena Dr. Edward Teller to ask him about self-interested corporation and military support of the continuation of testing."

On the other end of the spectrum were the likes of Fulton Lewis Jr., who is excerpted here from the *Washington Report*:

> "Dr. Linus Pauling, Cal Tech professor who is ring leader in the big noise against further atomic tests, now being staged with a congressional committee, deserves identification... The professor's record shows him to be one of the most prolific Communist front joiners in the business; the suspect petitions he has signed are almost uncountable."

Amidst it all, Pauling gave a number of speeches and participated in several demonstrations while he awaited his hearing. On July 9, a couple of weeks after his first appearance before the SISS, Pauling co-led a peace march with Ava Helen through downtown Los Angeles. When the march reached its destination, Pauling gave a speech that addressed many of the topics that would characterize his public talks throughout

the rest of the summer. He discussed in particular the apocalyptic consequences of a nuclear war, the development of nuclear weapons, and
the need for international bomb test agreements, as well as cooperation
with the Soviet Union.

Pauling also continued to speak out about the dangers of fallout and
the need for total nuclear disarmament, and he attempted to emphasize
the relationship between peace and freedom, a topic made even more
relevant by his impending battle with the SISS. Though the subject did
not tend to overwhelm his general message, he took great care to mention his difficulties with Senator Dodd whenever the chance arose. And
when speaking of Dodd, the intensity of their association often showed
through Pauling's choice of words. For instance, during his speech after
the march through downtown Los Angeles, Pauling criticized the "mad-
ness" and "evil" of Dodd's pro-nuclear viewpoints.

Once the hearing was postponed to October, Pauling and Wirin
attempted, unsuccessfully, to push the proceedings further back, to
November. Thwarted in this effort and pressed for time, Pauling worked
to address some of the committee's pending requests. He directed Wirin
to send committee counsel Sourwine a list of the names and addresses
that he had sent petitions to, as agreed, and took measures to consolidate
all signatures contained in the petition for presentation. He furthermore
had a book bound with 438 pages of names from U.S. petition signers and assembled all the signatures to be handed over for photostatic
copying.

Pauling questioned the subcommittee's authority to demand signatures from other countries however, and spent some time discussing this
aspect of their demand with Wirin. He likewise harbored many similar
concerns about the legitimacy of the subcommittee's authority regarding
other requests, and began taking substantive legal measures to address
them.

Pauling was planning to refuse a part of the committee's request
with finality, and understood well the potential consequences for
the crime of contempt. He was determined therefore to resolve the
matter before his scheduled appearance in October. He first sought a

declaratory judgment from a federal District Court stating that he need not submit to all of the committee's demands. After losing the case, he appeared before the Court of Appeals for the District of Columbia, which also ruled against him.

Legal analyses following Pauling's ordeal surmised that the decisions handed down against Pauling in both cases were likely not made because of their merit, but rather because the controversy between Pauling and Dodd was clouding the process of adjudication. Pauling eventually appealed to the U.S. Supreme Court, but his case was pending up to the day of his appearance before the SISS. Though Pauling's attempt to prevent his full hearing before the SISS was carried out with zeal, it appears that the courts had no desire to intervene in the matter before appropriate measures had been taken by all parties.

After months of maneuvering and preparation, Pauling arrived in Washington, D.C. in October, several days before the start of his hearing. At 11:00 P.M. on October 10th, the night before his scheduled SISS appearance, Pauling was once more served with a subpoena outside of the Congressional Hotel in Washington, D.C. The document commanded Pauling to bring with him all signatures to the petition, and all letters by which such signatures were transmitted to or received by him. With Abraham Wirin by his side and a bound list of petition signatures in hand, Pauling made ready the final preparations for his case.

*

"Until Supreme Court of United States decides Pauling v. Eastland, Dodd and others pending before it, Dr. Pauling will not appear before your subcommittee and bring with him the documents ordered. Respectfully request that committee postpone hearing date until October 26 on assumption court will act on pending petition for writ of certiorari either October 17 or October 24. Your attention is called to fact that McClellan committee took similar action when similar case was pending in U.S. Supreme Court."

— A.L. Wirin, Telegram to Senator Thomas J. Dodd,
October 10, 1960

The telegram excerpted above was sent from Pauling's counsel to Senator Dodd's office the afternoon before Pauling's hearing was set to take place. As might be expected, it was not well-received. Indeed, the action was apparently so insulting that Senator Dodd felt compelled to address the matter during his opening statement at Pauling's hearing, which did in fact take place the following morning. According to Dodd, the move was a "deliberately contemptuous challenge to the authority of the Senate of the United States." The telegram, as Dodd disdainfully relayed, was followed several hours later by a press conference where Pauling declared that he would not appear at the hearing as scheduled.

Similar complaints consumed a substantial portion of the senator's opening statement, the sum of which was a general reprimand of Pauling and his counsel. The scolding, like much of the affair, seemed in tone and language to be directed not at Pauling and Wirin, but rather at the wider audience for the day's events. In this vein, Dodd conveyed in carefully scripted detail his reaction to Pauling and Wirin's last-ditch effort to delay the hearing. According to Dodd, the subpoena that Pauling received at his hotel the night before had resulted directly from this turn of events.

And so began Linus Pauling's hearing, as scheduled, at 10:35 A.M. in the New Senate Office building on Tuesday, October 11, 1960. Present on the SISS were Senator and chairman Thomas J. Dodd, chief counsel Jules G. Sourwine, director of research Benjamin Mandel, and chief investigator Frank Schroeder. Other Senators that had been present for Pauling's June hearings were noticeably absent; a recitation of Nebraska Senator Roman Hruska's telegram, which apologized for his absence, was the first order of business for the day. From there, addressing the various media elements present, Dodd ordered that there was to be no recording of the proceedings and that only a two-minute time window for photos would be allowed.

After voicing his frustration with Pauling concerning the events leading up to the hearing, Dodd made sure to clarify once again that Pauling was not on trial. According to Dodd, the purpose of the occasion was simply to "secure information, on the record, under oath, that would be helpful to the Congress in the discharge of its legislative duties." Those present were meant to understand that the hearing was necessary for proper legislation. Dodd also took pains to address perceptions by some that the dispute with Pauling had somehow jeopardized the right to petition, sharing his opinion that the subcommittee had violated no rule of law. Permeating the atmosphere of the day, however, was the threat of a contempt of Congress charge, an offense that could lead to imprisonment should Pauling fail to comply with each of the subcommittee's requests.

After reading a part of his drafted opening statement, Dodd proposed that he not finish, but instead insert the full opening statement into the official record following the testimony. After a few objections and some discussion, Pauling and Wirin accepted the action and the proceedings were allowed to continue. Chief counsel Sourwine then introduced a number of pleadings which were ordered into the appendix of the official record. Afterward, Dodd and Pauling discussed turning over the bound signatures from the United Nations bomb test petition for photostatic copying, and agreed to have them added to the hearing record.

Once these details had been cleared up, Dodd finally came to the question that everyone had been waiting for: Had Pauling brought with him all letters of transmittal? Would he provide the subcommittee with the names of those who had helped him to compile the petition?

After a series of comments, wherein Pauling requested assurance that he would be given the chance to clarify his decision and respond to the remarks of Dodd's opening statement, he answered the long-awaited question:

> "I have not brought with me the documents, namely, 'the letters of transmittal by which or in connection with which such signatures were transported to you or received by you,' for the reasons that I presented

in detail at my hearing on the 21st of June, when I was asked if I would tell the subcommittee who had gathered the signatures, and how many signatures each person had gathered. And I replied that I would not do this because I was unwilling to subject people who are innocent of any wrongdoing to the reprisals of this subcommittee, that my conscience would not permit me to sacrifice these innocent people, some of whom had been without doubt led into this activity by their respect for me, in order to protect myself."

Faced with imprisonment Pauling had reacted with defiance, his brave response bringing with it an end to months of suspenseful speculation. After a brief pause, chairman Dodd simply replied, "Very well."

For much of the remainder of the hearing — which extended past a lunch break into the afternoon — chief counsel Sourwine took control of the proceedings, meticulously questioning Pauling about his association with 34 organizations and 25 individuals that had come under the suspicion of the subcommittee. Pauling replied to most inquiries with characteristic wit, and Dodd was often forced to gavel down the frequent bouts of laughter that resulted from Pauling's answers. Though the hearing was by most measurements stern and rather grim, one gains the impression that Pauling was enjoying himself at least a little bit during his testimony, perhaps relieved that Dodd had blinked on the issue of Congressional contempt.

Sourwine's interrogation continued until the very last minutes of the hearing. The final line of questioning was directed at Pauling's involvement with the Pugwash conferences, which, like most individuals and organizations mentioned during the proceedings, were accused of acting as a propaganda tool suffering from undue Communist influence. Pauling was given a chance to address these final accusations and then the hearing was ended, quite suddenly, after Sourwine stated that he had no more questions for the witness. After roughly six and a half hours, Linus Pauling was officially excused from his subpoena and the subcommittee was adjourned.

In hindsight, the merit of the October hearing is brought into question by a simple comparison of the purported purpose and final results. By most measures, the subcommittee failed to produce any new or relevant information for its own use during its questioning. Instead, the hearing seems to have been used to let the press and others know what the subcommittee found out about Pauling's contacts and associations during months spent dredging through his past. It likewise appears that, well before the trial, Dodd understood that public sentiment and media exposure would expose him to political risk were he to charge Pauling with contempt. But Dodd was loath to let Pauling escape unscathed, and as a result the hearing proceeded as scheduled.

So Pauling stood up to the subcommittee, refused to release information he felt inappropriate, and escaped a contempt charge. But what of other damages to his reputation in the public sphere? According to Harry Kalven Jr., a law professor from the University of Chicago who extensively researched aspects of the case, the results of the day were not so damaging as might have been supposed. In Kalven's words:

> "What emerges then as 'his long record of service to Communist causes' is that he favors repeal of the Smith and McCarran Acts; that he has doubts about the justice in the Rosenberg and Sobell cases; that he did not like the deportation of Hans Eisler or David Hyun, a Korean about whom he makes a particularly effective statement; that he favored abolition of the Attorney General's list; that he protested the treatment of the Hollywood Ten before the House Un-American Activities Committee in 1948; that he opposed the contempt citations against defense counsel in the Communist trials and the proceedings against the Jefferson School under the McCarran Act; that he thought the Hollywood Committee on Arts, Sciences, and Professions in which he, along with Mrs. Roosevelt, was active around 1948 was a useful idea; and that he has a deep concern about world peace which leads him to participate in many movements for it."

In short, the red-baiters had precious little to add to their dossier. And from Pauling's perspective, after months of planning, stress and legal maneuvering, the hearing went about as well as could have been imagined.

*

"The [SISS] hearings shade imperceptibly from the incompetent and absurd to the nasty and sinister; there is some temptation to say that they are simply ludicrous — objectionable, perhaps, as a waste of taxpayers' money and congressional time, but not on grounds more closely related to liberty."

— Harry Kalven Jr., 1960

Though most were happy to see a conclusive end to Pauling's ordeal, many were perplexed by the docile nature of its conclusion. Writing in the *New York Times*, Anthony Lewis echoed the impressions of many in noting that, "The purported epic struggle between Dr. Linus Pauling and the Senate Internal Security Subcommittee has ended not with a bang but a whimper. What was supposed to be a showdown hearing this week came to no showdown at all. Dr. Pauling was excused without any orders to produce the documents he had refused to produce."

While the hearing failed to fulfill the expectations of many, it was still subject to a degree of scandal and intrigue. In particular, Dodd's partially read opening statement became a major point of contention in the hearing's aftermath. Having initially objected, Pauling and his counsel, in the midst of the hearing, had allowed Dodd to insert his full written opening statement into the official record without reading it aloud. And as it happened, the opening statement that appeared in the record was very much at odds with the one that had been partially read aloud to those present at the hearing. A close analysis suggests that Dodd seems both to have read early parts of the statement selectively and to have strategically stopped reading the opening statement just before blanket accusations of Communist influence could be uttered in public.

The modified opening statement transformed the purported purpose of Pauling's hearing after the fact. Though Pauling was eventually given an opportunity to insert his objections into the official record — he was afforded space for paragraphs no greater in length than those which were being commented on — Dodd's full statement was made available to press elements outside of the courtroom during the first part of the hearing. Not only was Pauling charged with Communist affiliation in the unabridged statement, but Dodd's written tone, reasonable if stern at the trial, took a turn toward the austere. One extract reads,

> "This committee has a continuing and longstanding interest in all phases of Communist propaganda, because as a part of our legislative duties we must constantly ask ourselves the question: What Communist propaganda activities are being engaged in, or are having an impact upon the internal security of this country, which can be countered either wholly or in part, by legislation? The Communists are constantly developing new techniques and revising old ones. We cannot, at any time, assume that legislation already on the books is giving our country the greatest possible protection from the Communist conspiracy. We must be constantly on the alert to uncover as much as we can of Communist activities and Communist techniques, and then apply ourselves to determine whether what we have found justifies recommendations to the Senate for proposed new legislation."

Dodd's complete opening statement was probably the most outlandish addition to the hearing's official record, but other controversial items were also inserted following (and possibly resulting from) Pauling's vehement post-hearing objections. Though most of the appendices following the official testimonies for the October hearing are mundane replications of court documents, petition signatures and pre-hearing correspondence, a special final report was included in the last few pages of the full official record. Compiled following the hearing, the report takes all aspects of recorded testimony into consideration and provides its own final analysis.

Given that certain section titles of the report are titled "Dr. Pauling's role in the international Communist 'peace' offensive," and "The Communists come to the defense of Dr. Pauling," one might correctly assume that the tenor of the document was hostile toward Pauling's positions.

The conclusion of the report synthesizes the final opinion of Dodd and his associates — an opinion reached after months of sour, drawn-out dealings with the famed chemist:

> "The subcommittee believes that study should be given to the possibil-
> ity of legislation which will make it more difficult for the Communists,
> and those who collaborate with the Communists, to abuse the right of
> petition by utilizing it for their own subversive ends... The subcommit-
> tee therefore recommends that hearings be held for the purpose of
> examining various legislative possibilities in this area."

Pauling seems not to have yielded any unnecessary ground throughout the hearings, but it is difficult to declare a clear winner in the wake of the conflict. On one hand, Pauling was not forced to comply with the committee's request to hand over his letters of transmittal, but the committee, in turn, was never forced to acknowledge or admit that its questioning was improper. Likewise, by choosing to not pursue contempt proceedings, the committee was able to avoid a very dangerous constitutional showdown, but was still able to hint at the full extent of its wrath to any future would-be challengers.

Pauling freely discussed his fight with the SISS following the hearings — and strongly considered writing a book about it — frequently framing the experience as an attempt to suppress his constitutional rights. At a speech delivered one month after his hearing to the Colorado Chapter of the American Civil Liberties Union, Pauling provided an indication of the feelings of injustice that lingered within him:

> "I believe that Senator Dodd has the right to advocate that a change
> in our government's policy be made, but that it is a misuse of his pow-
> ers as Vice-Chairman of the International Security Subcommittee of
> the U.S. Senate to harass me and to attempt to suppress me in the
> exercise of my right as an American citizen to work for international
> agreements for cessation of nuclear tests and ultimately for univer-
> sal and total disarmament, with the best possible system of control

and inspection. Let us demand that Congress abolish this committee, which is a disgrace to the Congress of the United States, to the nation itself, and to the American people."

Clearly Pauling felt that he had been subjected to unjust and unfair conduct, and saw that no substantive recourse was being made available to redress it. For some months, his work and speeches following the October hearing were laden with mention of Dodd and "the evil way in which this subcommittee is misusing its powers and subverting the constitution."

By several accounts Pauling took the proceedings very personally, and the resentment that he harbored was not slaked by simple public condemnation of Dodd and the subcommittee. Rather, during the year after his hearing, Pauling launched five libel lawsuits against several newspapers and political organizations that published what he considered to be defamatory statements. Following a dispute with founder Norman Cousins, Pauling also discontinued his association with the Committee for a Sane Nuclear Policy, and was even on the verge of suing the *Bulletin of the Atomic Scientists*, a magazine that had been nurtured by Einstein's Emergency Committee of Atomic Scientists. His participation in the Pugwash conferences also ended for a time (despite his unwavering defense of them during the hearing), because organizers continued to invite individuals whom Pauling no longer viewed favorably.

In the final analysis then, after months of bitter altercation, Senator Dodd and his committee had managed a feat that few others had accomplished — the unsettling of Pauling's nerve. While at first blush it appeared that he had ended his clash with the SISS as a hero of the peace community, Pauling's characteristic pride and resolute constitution did not make it through unscathed.

12 Unitarianism

Chapter

By Will Clark

Stephen Fritchman (left) with Pauling in La Jolla, California, 1969.

"In Unitarianism I have found a religion without dogma: A growing, changing, open-minded willingness to learn and, above all, to work."
— Ava Helen Pauling, "Why I am a Unitarian," September 18, 1977

Though both Linus and Ava Helen Pauling were avowed atheists, they did maintain a long and friendly relationship with the Unitarian Church. Key to this relationship was their connection with Stephen Fritchman (1902–1981), a Reverend of the Unitarian Church of Los Angeles and an influential member of several major peace groups. The Paulings first encountered Fritchman in 1945, through their mutual association with the Hollywood Council of the Arts, Sciences and Professions. They became more closely associated in 1949, through their participation in a memorial service for Pauling's trusted physician, Thomas Addis. Fritchman led the respectful and dignified ceremony, one that left the Paulings feeling comforted, hopeful and, as it would happen, interested in the Unitarian Church.

Both Linus and Ava Helen had rejected organized religion at very young ages. Notions of an anthropomorphic god or salvation dependent on intercession from a third party were discarded in favor of a doctrine grounded in reason, ethics and morality. In various interviews over the years, both of the Paulings evinced similar motivations for joining the Unitarian church, but took great care to mention that a creed or means of salvation were not among them. To this end, Ava Helen once listed seven principles of Unitarianism, formulated by a Unitarian committee in 1930, to highlight the similarities of outlook between science, Unitarianism and her own personal beliefs. They included:

1. Use of the scientific method in approaching religion.
2. Rationality of the universe and progressive discovery of truth.
3. Humility and reverence toward vaster forces of the universe.
4. Conviction of infinite possibility of human progress.
5. Free exercise of intelligence in religion.

6. Conviction of self-sufficiency of humanity to solve its problems.
7. Sense of human brotherhood.

For his part, Linus Pauling consistently emphasized his independence within the peace movement, while also remaining appreciative of the connections that he sometimes made with "like-minded individuals" in various peace organizations. As with his participation in many other peace- and science-related activities, Pauling's involvement with the Unitarian church offered an opportunity to network, share ideas and push an agenda through public speaking.

When Dr. Pauling spoke of the value of certain faith-based institutions, he sometimes likened them to gaps that he felt science could not fill through its methodologies. In a speech crafted for the First Unitarian Church of Los Angeles in 1969, Pauling observed that "despite our increasing power over nature, the amount of suffering in the world remains very great... The major reason for the continued misuse of our power over nature, for the continued existence of a great amount of human suffering, is that the world is not operated on the basis of an accepted ethical principle." Though he was openly averse to supernaturalism, mysticism and dogmatism, Pauling was fundamentally a humanist who believed that both humanism and modern Unitarianism were rational philosophies that could be a source of good in the world.

In due course, the Paulings became active church members, and Linus was invited to join the Unitarian Service Committee, a "nonsectarian, voluntary agency whose purpose [was] to promote human welfare through service." The Paulings also came to greatly admire Rev. Fritchman's tolerance of diverse opinions, his respect for individualism, and what they often referred to as his "great social conscience."

Indeed, soon after the resolution of Linus Pauling's initial bout of passport problems, Fritchman himself was denied a passport for travel to Australia, where he was scheduled to speak at a Unitarian celebration. In a letter of protest to the State Department, Pauling wrote, "I have known

Mr. Fritchman for several years. I consider him to be a great man. He is one of the most honest, forthright, straightforward and high-principled men that I have ever known. He is an honor to the United States of America — the world would be a great world indeed if one percent of its people were comparable to Mr. Fritchman."

During the 1950s, the Unitarian church became embroiled in a controversy involving California loyalty oaths. A case condemning the group was brought to trial, but the church was eventually vindicated by a United States Supreme Court ruling. This event, among many others, did much to strengthen the bond between the Paulings and the church.

Above all, the Paulings found in Unitarianism a unique forum to fight against the grave problems facing the world. Their participation in the church was not often separated from fervent activism or an acknowledgment of difficult political realities. Rather, the connections that they made fit well into their condemnation of war, nuclear proliferation, militarism and inequality.

13 An Honorary Diploma from Washington High School

By Madeline Hoag

Pauling holding his honorary high school degree, 1962.

"My early education in Oregon was outstandingly good. My father, who died when I was nine years old, had recognized that I had an unusual interest in reading and in obtaining knowledge. It is my opinion that the many excellent teachers I had contributed very much to fostering this interest. I remember in particular several teachers at Washington High: Miss Pauline Geballe, who taught me my first course in science;

William Greene, who taught me my first course in chemistry and gave me special instruction during the following year; and my several teachers in mathematics. I feel that my early education in Oregon was outstandingly good."

— Linus Pauling, 1962

Although Linus Pauling received honorary doctorates from some of the most prestigious universities in the world, it was not until 1962 that he received his 20th degree, a high school diploma, from Washington High School in Portland, Oregon. The 45-year gap between senior year and diploma was due to a regulation that forbade the young Pauling from taking two courses in American history during the same term, ultimately preventing him from graduating in June 1917. At that time however, Oregon Agricultural College accepted students who had not graduated from high school, so off Pauling went to Corvallis and, eventually, a long and storied life in academia.

The idea to award an honorary high school degree to Pauling was initiated by Jerry Ross, a journalist for *The Washingtonian* school newspaper. Ross had attended a press conference given by Pauling on May 4, 1962 at Portland State College in which he talked about his childhood and mentioned that he was a former student at the high school but not a graduate. Ross reported this information to Principal Harold York, who then wrote to Pauling about the matter. In his letter, York exclaimed, "We have taken steps to correct this embarrassing action. Washington High School is intensely proud of the fact that you are numbered among its most illustrious former students."

Along with the note, York enclosed the honorary diploma, an issue of *The Washingtonian*, a copy of Pauling's high school transcript, and a commencement announcement coupled with tickets to the exercises. York made sure to point out that the old Washington High School had burned to the ground in 1922 — it was rebuilt as a brick building in 1924 — and most of its records were destroyed, which may be the reason why Pauling's transcript appeared to be incomplete.

With the commencement tickets, York extended an invitation to Pauling to attend, as an honored guest, that year's exercises, to be held on Wednesday, June 13, 1962 in the Benson High School auditorium. He also asked if Pauling would take a picture of himself looking at his diploma for publication in the school paper.

In response Pauling expressed his appreciation for the honor, especially as he was the first and only honorary graduate of Washington High School. "I am happy to be a member of the graduating class of 1962," he wrote, "and I send my regards and congratulations to the other members of the graduating class." He then apologized for being unable to attend the ceremony due to a previous speaking engagement but noted that he would never forget the school and his many exceptional teachers, and promised that he would pay a visit at a later date.

Following Pauling's acceptance of the honorary degree, *The Washingtonian* published an article with the headline, "Dr. Pauling is 1962 Grad! Worldly Physicist Gets Only Honorary Washington Degree." Reporter Ross cheekily opined in the article that the high school diploma seemed justifiable since Pauling had shown that he was certainly capable of post-high school caliber work.

The *New York Times* also published a brief article of its own, emphasizing that Pauling had "finally received a high school diploma." More thoughtfully, the *Meriden (Connecticut) Record* commented on the story with an editorial titled "Belated Diploma," published July 2, 1962. In it, the newspaper suggested that

> "the lessons may not have been the ones in the prescribed high school textbooks, and then again they may. But the concern for people in society, their freedom, safety and well-being has been Dr. Pauling's distinguishing characteristic... He has made the rest of us think, and forced us to face our own responsibilities a little more squarely in this matter, too. Clearly, this is social studies at its best, considerably above the high school level. Dr. Pauling did it the hard way, but he has surely earned his diploma."

*

The history of Washington High School dates back to 1906, when it first opened its doors to students. The school was originally called East High School, but changed its name to Washington three years later. Located at SE 14th and Stark, the school merged with Monroe High School, an all-girls school, in 1977.

Washington-Monroe High closed in May 1981 due to declining enrollment, after which time the building was used for administrative purposes with hopes of it one day becoming a community center, should additional funding become available. In its later incarnations, the old school also hosted a daycare center, a continuation curriculum for pregnant girls, a "vocational program for Indian youth," and learning opportunities for a segment of Portland's special education students.

Portland Parks purchased Washington High School's west field in 2004 and a few years later tore down the old gym and cafeteria. Once the buildings had been demolished, it took another five years to secure the funds to draw up plans for a new facility. In an article published in *The Portland Mercury* on August 20, 2009, neighborhood association member Kina Voelz described a collective dream for the space, which might entail everything from a photography studio to a rooftop garden on top of the new building. "There's been all kinds of wish-list discussions about this in the neighborhood for years," Voelz said.

Some of the spirit of this vision was realized in 2014 when Revolution Hall opened. A music performance space carved out of the former WHS auditorium, Revolution Hall features 850 seats, a rooftop deck, two bars, and artwork depicting Pauling juggling four oranges.

*

The 1959 edition of the *Lens*, Washington High School's yearbook, was dedicated to Linus Pauling, describing the professor as one who "personifies our highest ideals." Published three years before Pauling's honorary diploma, the first page of the volume featured a head shot of Pauling along with a brief write-up about why he had been chosen for the dedication. The yearbook further mentioned that Pauling learned the fundamentals of chemistry at Wa-Hi under the direction of William V. Green, a Harvard graduate and former *Lens* adviser.

Seven years later, Pauling made good on his promise of one day returning to Washington High School when he attended its 60th anniversary celebration on February 22, 1966. Former administrators, faculty and students were welcomed back for this event, which was held in the school auditorium. "Memories of School Days" was the theme of the evening, which included messages from alums and small-group conversations organized by year of graduation.

Pauling delivered a talk at the event, speaking on the "foolhardy aspect of brinkmanship in a nuclear age and implor[ing] young people to think as individuals." He also met with physics and science seminar classes, signed autographs, and was photographed alongside Principal York.

Chapter 14 — Receiving the Nobel Peace Prize

By Andy Hahn, Jindan Chen and Michael Mehringer

Celebrating the Nobel Peace Prize announcement, October 1963.

Like many other Nobel Prize winners, Linus Pauling discovered that he had been awarded the Peace Prize in a dramatic way. The news was announced on October 10, 1963, while Pauling was at his Big Sur ranch — an intentionally secluded space lacking a telephone, to say nothing of a television. He, Ava Helen and some friends had already planned on celebrating that morning, as October 10th would also mark the formal beginning of the Limited Test Ban Treaty, which put an end to above-ground nuclear tests among the world's major nuclear powers. As he wrote in his research notebook:

"Ava Helen and I had come to the ranch with Mr. and Mrs. Clifford Durr. We had bought a bottle of champagne, which we planned to drink to celebrate the treaty. At 8:15 A.M., as we were sitting down to breakfast, the forest ranger, Ralph Haskin, came to the cabin. He said that Linda had telephoned and had asked that Ava Helen and I both come to the ranger station and telephone her. I asked if he knew what was the matter, and he said that he thought that it wasn't serious. (Linda had told him and asked him not to tell us.) We finished breakfast, drove to the station, and telephoned Linda. She said that I had been awarded the Nobel Peace Prize for 1962. (She first asked me if I had heard the news. I said no.) I spent most of the day at the station, answering the telephone and giving interviews. We forgot to open the champagne. On 11 October, we drove to Carmel. Ralph Atkinson had champagne at hand. It's our first celebration."

The fact that Pauling received the 1962 Prize in 1963 is extremely telling. As Pauling wrote in a confidential note to self:

"On the morning of Tuesday 13 Nov., Gunnar Jahn [then chair of the Norwegian Nobel Committee] telephoned me at the Bristol Hotel, Oslo, and asked us to come to his office at 11 A.M. There he said to Ava Helen and me, in the presence of his secretary, Mrs. Elna Poppe, 'I tried to get the Committee...to award the Nobel Peace Prize for 1962 to you; I think you are the most outstanding peace worker in the world. But only one of the four would agree with me. I then said to them, 'If you won't give it to Pauling, there won't be any Peace Prize this year.'"

And indeed there was not.

Jahn's conflict with his colleagues was symbolic of the differing attitudes with which news of Pauling's Peace Prize was greeted, especially in America. While the public and many of Pauling's friends showered him with congratulatory letters and telegraphs, pro-nuclear scientists, much of the mainstream media, and official agents of the U.S. government were generally unhappy about the accolade.

Perhaps most famously, on October 25, 1963, *Life* magazine published an editorial titled "A Weird Insult from Norway," which, as one might imagine, criticized the Nobel committee's decision. The critique specifically attacked Pauling's Prize from two directions. First, the editors pointed out that the recognition of Pauling's peace work by the Norwegian committee was, in effect, a condemnation of contemporary research on nuclear science. The magazine argued that if efforts to ban nuclear tests were deemed worthy of respect, then efforts to promote nuclear research were conversely discredited. By this logic, Pauling's Nobel Peace Prize was presumed to be an insult to other scientists engaged in nuclear weapons research.

Second, the *Life* editorial sought to undermine Pauling's importance to the nuclear disarmament movement. The magazine stressed that the real reason why the Partial Test Ban Treaty came into being was not because Pauling's famous 1958 appeal had finally changed the minds of governments, but rather because President Kennedy's firm stance against the construction of missile bases in Cuba during October of the previous year had, to a large degree, helped shape sentiment in favor of disarmament on a global scale.

While Pauling received many letters of support from those who were outraged by the editorial, few were quite so colorful as that penned by his friend Ernst Scharrer. At the time a faculty member at the Albert Einstein College of Medicine, Scharrer began by dismissing as folly the logic behind *Life*'s critique. From there, Scharrer compared the editors' published opinion to Adolf Hitler's response to Carl von Ossietzky's 1935 Peace Prize. As Hitler secretly began rearming Germany, in the process ignoring the 1919 Treaty of Versailles, Ossietzky revealed the news to the world by publishing details of the militarization. When this effort won Ossietzky the Nobel Peace Prize, Hitler declared that, henceforth, German citizens were forbidden to accept Nobel prizes. Though of lesser consequence, Scharrer's point was that *Life*'s critique was similarly unjustified, partisan and petty.

But what explains the divergent views of Pauling's peace efforts? To answer this, it helps to go back to the central questions of the argument over nuclear testing. Opposing viewpoints on these questions were, to a degree, summed up in a televised debate between Pauling and Edward Teller titled "Fallout and Disarmament" and broadcast live on San Francisco's KQED-TV on February 20, 1958. Teller spoke for the pro-nuclear camp in first explicitly stating that, as with Pauling, peace was his goal. The focus of the controversy, then, was how best to bring about a world in peace, the conflict centering on whether or not the process should involve nuclear weapons.

The still somewhat unknown side effects associated with the manufacture and testing of nuclear weapons also made this an issue of pitched debate. It was known that nuclear bombs could crush islands into dust and spread them into the atmosphere over the course of a few seconds. What's more, rare, toxic elements were also clearly created alongside the release of large amounts of radiation. But no one knew exactly what might happen to our bodies when exposed to regular, if lower, levels of these atomic-era materials.

Faced with this uncertainty, the two sides came into conflict on the question of what the development of nuclear weapons might bring to American society. Teller thought them beneficial in that the ability to manufacture these massive weapons meant that the U.S. could match and possibly overcome the Soviet Union's military strength. It was this balance between the two military superpowers, Teller claimed, that would guarantee peace. In the absence of this dynamic, war was presumed to be inevitable, with the side that failed to develop a matching nuclear capacity finding itself at a distinct disadvantage. For Teller, the arms race necessitated the testing of nuclear weapons. The real stake was military strength — peace was based on force. In addition, and of major consequence to his position, Teller maintained a very optimistic view on the health effects of prolonged exposure to fallout levels of radiation.

He even pointed out the possibility that increased mutations resulting from fallout could be viewed positively, as a source for enhanced evolution of species.

Pauling couldn't have disagreed more vehemently. He thought the construction of nuclear weapons to be a bane of world society and emphasized the environmental and health costs imposed by their development. Morality was another of Pauling's tools with which to attack Teller's arguments. If it was understood that developing weapons to strengthen national security came at the cost of decreases in public health and environmental stability, the effort, even if well-intentioned, was morally corrupt and ought to be brought to a close as soon as possible. Furthermore, as a fundamental principle, world conflicts should be settled at the negotiating table instead of the battlefield. Pauling also expended much energy in compiling evidence on the ill effects of increased environmental radiation. One example that he often cited was the increase in the incidence of children born with birth defects after World War II and its nuclear conclusion. By directing his audience's attentions to the impact of atomic gamesmanship on future generations, Pauling stressed the seriousness of the issue and re-emphasized the morality of his argument.

The crux of the Pauling-Teller debate was still in play by the time Pauling received his Peace Prize. While *Life* magazine was criticizing the decision of Norway's Nobel committee to reward Pauling's peace work, his emphasis on moral action was being enthusiastically supported by his friends and many others in the public arena. In particular, Pauling's concern over the harmful health effects of atmospheric radiation on future generations gained a lot of traction with the public. This positive response was reflected in many letters of congratulation from ordinary people who wholeheartedly endorsed Pauling's appeals. Typical was an October 13, 1963 letter, written by a widow with two boys:

"To me you have been vindicated in the eyes of the world. These stupid, loud-mouthed patriots, as they consider themselves, should have to eat their words. I am not a college educated person, and I do not pretend to know what the ultimate outcome of this testing program would be, but I have read enough to make me very fearful as you are. I think we all should consider the future generations — not just ourselves, as you did. But few would be as brave and heroic as you, and would 'stick our necks out' as you did. You are a truly great American and a great humanitarian, which is more important! Someday people will speak of you as the great man you really are. I feel so relieved that you have won this prize, as I have been very bitter over the criticism of you. I have resented it so much, but now I feel people will change in their opinion of you...if they don't, these few 'screwballs' you should not care. Most of us are as happy as if we won that prize ourselves. I know I am! Usually it seems, they wait until you die to relent and say a person is truly great and deserves the highest honor. So I feel so grateful that this was done while you still can appreciate the fact that you are considered by many a hero, if there ever was one!"

*

The controversy stirred up by *Life*'s October 25th editorial continued throughout November as other publications began to weigh in.

In its November 2nd opinion piece, *The Nation* attacked *Life*'s — and much of the mainstream media's — response to Pauling's Nobel win as superficial. In so doing, the magazine noted how the press normally "flips" when Nobel Prize winners are announced, excitement generated only because of the Prize itself and often absent during the years of hard work leading up to a Laureate's receipt of their Prize. Pauling's case had been unusual though, in that the press had flipped in a different way: Pauling was evidently to be criticized because, through his activism, he had overstepped his boundaries as a chemist as well as those of the usual "verbal peacemaking." Despite this, *The Nation* editors remained confident that Pauling would "take it all with equanimity."

This notice of support was buttressed by another article in the same issue that was written by Philip Siekevitz, a professor of Biochemistry at Rockefeller University. Titled "Scientists and the Public Weal,"

Siekevitz' piece emphasized scientists' obligation to social responsibility, a duty as important as the commitment to truth in the laboratory. Compellingly, Siekevitz wrote that scientists like "Teller and Pauling are damned by their colleagues for their politicking not so much because of the side of the political fence each happens to be on, but simply because they talk about scientific matters in a political way." By stepping out in this manner, Siekevitz argued that scientists could help balance the more destructive aspects of scientific progress that had been evident in the first half of the 20th century.

An editorial in the November issue of *Liberation* both lauded and critiqued Pauling's Peace Prize win. It first noted that several American Nobel Peace Laureates from the past had "played prominent parts in some war" and that Pauling was a "refreshing variant." In spite of this, the editorial more critically offered that Pauling "is not a pacifist; nor is he a crusader for unilateral disarmament." The specific source of the piece's misgivings was Pauling's actions at the Oxford Conference in January 1963, where he seemed "insensitive to the fact that there may be honest differences of opinion, not dictated by expediency or bureaucratic biases, as to how peace and unity in the work for peace can in fact be promoted." Pauling, though, could still contribute to "the frankest and closest discussion among all peace workers of the issues involved (including those on which we differ), of the nature of the goal, and of the way by which it can be reached."

The European press picked up on the controversy as well. In West Germany, *Das Gewissen* — one of several media outlets in the country to comment on Pauling's Nobel — explained to its readers how Pauling's activism could be so controversial in America. In Spain, Julian Marias wrote in the magazine *Gaceta Illustrada* that he harbored some reservations about Pauling's work, but was still supportive since Pauling had previously warned him about the harmful effects of nuclear fallout. Many others in the European press more simply announced Pauling's win and impending travels to receive the award.

Individuals were also able to weigh in on the controversy, including readers who wrote in to *Life*. Some, like Mrs. Norman Yves of Temple City, California, approved of the magazine's stance, writing that she was "glad that LIFE had the courage and the integrity to inform the public of the real Linus Pauling" and congratulating its editors "for publishing the truth." Others, like Robert F. Adam of Emporia, Kansas, appeared indifferent to the controversy, noting "I'm glad that I live in this country where a man can voice his opinion against policies of the government and can accept recognition for these opinions from another country" — not an option, he stressed, available in Communist countries.

Still others, like Thomas L. Allen of Davis, California, took offense to the portrayal of Pauling by *Life*, and sought to correct it. Allen drew specific attention to the 1958 petition that Pauling had written with Edward Condon and Barry Commoner opposing the testing of nuclear weapons, pointing out that it "does not demand the 'uninspected stoppage of all nuclear tests;' it calls for an 'international agreement to stop the testing of all nuclear weapons. The question of inspection is not mentioned in the petition."

In an angry letter of his own to Henry R. Luce, the editor of *Life*, Pauling agreed with Allen:

"As you know, Life Magazine was guilty of an outright lie, in its perfectly clear statement, in its editorial on 25 October 1963, that I had demanded the uninspected stoppage of all nuclear tests. If you and your magazine had any standards of morality, you would have corrected and retracted this statement, and have apologized for it. All that your magazine has done is to make a statement of acquiescence to a correction in a letter from a correspondent, with an added quibbling statement apparently designed to confuse the reader."

(In this, Pauling was referring to *Life*'s reply to reader Allen: "It is true that Dr. Pauling's petition did not demand an 'uninspected' test ban. But it did call for 'immediate action' without any reference to inspection, a significant omission because an enforceable inspection system was then the main issue separating the U.S. and Russia.")

Colleague Victor Herbert advised Pauling to sue *Life* "on the grounds that the wording in the Editorial is a deliberate distortion of fact whose wording is calculated to imply you are an extreme leftist."

Pauling's response was sympathetic but restrained:

"I registered a strong complaint with Henry Luce, and asked for a retraction. The retraction was made, in a grudging and quibbling way, in the letters to the editors section. I myself feel that the editorial involved a deliberate distortion — I would use even stronger words; but I am not sure that it would be worthwhile to file a libel suit."

With all the heaviness surrounding the controversy, F.A. Cotton of MIT was able to bring in a bit of needed levity, writing to Pauling:

"May I take this opportunity to express my pleasure at your receiving the Nobel Peace Prize and my admiration for your courageous and important work in alerting humanity to the dangers of nuclear testing. I might also say that for very good positive reasons, as well as to further increase Mr. Luce's ludicrous frustration, I hope you will soon receive the Nobel Prize in Medicine."

Surrounded by attention from the press, Pauling took the opportunity to conduct a few interviews despite his busy schedule. Notably, reporter Arthur Herzog from the *New York Times* had approached Pauling about a one-on-one for an upcoming article that would appear in the paper's magazine in mid-December. Herzog requested that Pauling come to New York, but Pauling had to decline. To make things easier, Herzog said he could come to Pasadena to interview him on November 12th and 13th.

While Pauling agreed, he did so reluctantly and, as it turned out, this reluctance was justified. Just a few days before leaving for Norway, Pauling sent a letter to *Times* editor Arthur Hayes Sulzberger stating that he had "received a telegram from [Herzog] saying that the New York Times has rejected the article that it had commissioned him to do." He continued, "I feel that the New York Times owes me an apology for having, through what seems to be misrepresentation, caused me needlessly to interrupt the very important work upon which I was engaged."

Pauling ended his letter by challenging the *Times* to live up to its reputation:

> "I feel that it is important that the people of the United States have a newspaper upon which they can rely. My opinion of the New York Times is necessarily determined in considerable part by the actions of the New York Times with respect to me — that is, with respect to matters about which I have personal knowledge. I feel that I may be forced to make some seriously critical statements about the New York Times to the American people."

To avoid further conflict, Sulzberger's son, Arthur Ochs Sulzberger, quickly responded. In so doing, the younger Sulzberger's carefully worded apology referred Pauling to a recent editorial that "is a true expression of the high esteem in which you are held." The editorial tried to balance what direct criticism it contained of Pauling with comments from those who held him in high regard, saying that Pauling "is of course one of the world's great chemists; and when his achievements in that field were recognized by the award of his first Nobel Prize some years ago, no one could have taken exception." That said,

> "He has not always been wise in his choice of tactics and he has sometimes been reckless. But Gunnar Jahn, chairman of the Norwegian parliamentary committee that chose him for the prize, could well ask yesterday if the nuclear test-ban treaty would have yet been achieved 'if there had been no responsible scientist who tirelessly, unflinchingly, year in, year out, had impressed on the authorities and on the general public the real menace of nuclear tests?' That scientist was Dr. Pauling. His courage in running against the crowd is now being recognized."

The letter must have been enough for Pauling as he made sure that his *Times* subscription would continue when he moved to Santa Barbara early the next year.

*

Separate from the stress of managing his public image, Pauling's response to the flood of supportive letters that he received was one of gratitude and recognition that he would not have won the Prize without the work of many others around the world. In the November issue of *War/Peace Report*, Pauling pointed out, "I share this prize with the major part of the scientific community and especially with Bertrand Russell."

And as more and more letters came in, he began to expand his recognitions even further. In response to a telegram from the Canon of St. Paul's Cathedral in London, L. John Collins, Pauling agreed that the Peace Prize "was a 'richly deserved honor,' as you say. In fact, I think that there was a good bit of luck involved — perhaps that the Norwegian Nobel Committee felt that I was a good representative of the organized and disorganized nuclear disarmament movement."

The award was, in addition, "having a great influence on the peace workers of the whole world," as Pauling told the German theologian, Martin Niemöller. Ultimately, as Pauling would confess to the President of the Women's Union of Argentina, the Peace Prize was "an honor not earned by me alone, but rather also by you and your associates, and by other peace workers throughout the whole world."

Indeed, the attention that Pauling received from various organizations after winning the Peace Prize could only have made his indebtedness to other peace workers more apparent. Teas and dinners were held around Pasadena in Pauling's honor, and the American Civil Liberties Union feted him — alongside Leroy Johnson, the first African-American state senator in Georgia since 1871 — at its annual Bill of Rights Banquet in Los Angeles. Throughout, Pauling was surrounded by supporters and fellow peace advocates, assuring him that he was not alone in his fight.

Pauling also garnered celebratory attention from the peace community with the establishment of the American Peace Prize. In the October issue of the newsletter *Peace Concern*, published in Montpelier, Vermont, this new award was announced, along with a drive to match the $51,000 that came with the Nobel award. The American fund would be compiled through donations from individuals of one dollar each, and by the end of the year, bolstered by organizational assistance from the group Women for Peace, over 1,000 people had made donations of varying amounts for peace activities to be administered by the Paulings. For good measure, the writers of *Peace Concern* also noted that the fledgling award "does strengthen our feeling that the American peace movement can, through the American Peace Prize, avail itself of the pleasure of kicking in the face (nonviolently, now!) 'Life's' assertion that '...the eccentric Dr. Pauling and his weird politics have never been taken seriously by American opinion.'"

The American Nobel Memorial Foundation likewise emerged as an organization that wished to honor Pauling, but they quickly aroused his suspicion. Pauling was first approached by the executive chairman of the foundation, Jacques F. Ferrand, in a letter dated October 21, in which he reminded Pauling that he had attended the foundation's dinner in 1961 and requested that he join them again at their upcoming dinner, scheduled for New York City in connection with the 1964 World's Fair. Ferrand further proposed that, at this dinner, which would also be attended by Senators and other members of Congress, a "committee for nominations" be formed for Nobel Peace Prize candidates and recipients.

In just over a week, Pauling responded that he saw "no need for deliberation by such a Forum" since many people, including members of the Senate and Congress, could make nominations as individuals. Pauling also sent carbon copies of his letter to others in hopes of alerting them to the activities of the foundation. Among these correspondents was Albert Szent-Györgyi, the discoverer of vitamin C and winner of the 1937 Nobel Prize for Physiology. Pauling pointed out to Szent-Györgyi that the American Nobel Memorial Foundation claimed him as honorary chairman.

Szent-Györgyi replied that he was surprised but presumed "that they have asked my permission to use my name and it is quite possible that I consented. In any case, I have forgotten about it." What he did know about the foundation was that "the Nobel Committee, in Stockholm...disavowed all connection with the American Nobel Memorial Foundation," a detail that Pauling would confirm for himself in his communications with Gunnar Jahn.

For his part, Ferrand "was not only surprised but most disillusioned" by Pauling's response, and expressed a hope that the two of them could work out a way for the public to be more directly involved in selecting Peace Prize recipients through their representatives in government. Pauling maintained his position, reiterating his suspicions of the foundation's proposed actions and asking for copies of its legal documents, including its articles of incorporation. This last move seems to have silenced Ferrand, as no reply remains extant. And in April of the following year, New York State Attorney General Louis J. Lefkowitz brought charges against the American Nobel Memorial Foundation on behalf of the Swedish Embassy for using the Nobel name under misleading circumstances.

Amidst all of the accolades and controversy, Pauling still had to plan his trip, which he eventually decided to restrict to just Norway and Sweden. It is customary for Nobel recipients to refrain from giving any public talks between their nomination and acceptance, yet the Norwegian Student Association approached Pauling about speaking on the day of the Nobel ceremonies at a separate ceremony that would combine the celebration of the 15th anniversary of the Declaration of Human Rights with an anti-apartheid meeting.

August Schon, the director of the Nobel Committee of the Norwegian Parliament, had alerted Pauling to the possibility that he might be approached by the students and reminded him of the usual practice that Laureates not speak before accepting their award. Pauling's response to the Norwegian Student Association was much more straightforward: he was simply too busy, with a note to himself that he "shouldn't do anything else." Inundated by a flood of additional invitations, Pauling delegated the scheduling and planning of his post-acceptance speaking tour to Otto Bastiansen, a professor at Oslo University.

Near the end of November, just prior to leaving for Europe, the Paulings were able to get away to their ranch for some peace and quiet. While there, Pauling worked on his acceptance speech and his Nobel Lecture, "Science and Peace," and Ava Helen likewise drafted talks that she had been invited to give. Pauling had become so adept at writing his speeches that the draft that he left the ranch with, which he composed over six days, remained largely unaltered once he reached the podium in Oslo.

*

As congratulatory letters continued to pour into his office in early December 1963, the debate in the press over whether Pauling deserved the Nobel had begun to cool down. Meanwhile, closer to home, friends and colleagues hosted several events honoring the Paulings for their activism.

On December 1st, eight activist groups, including Women Strike for Peace, the American Friends Service Committee, SANE, and the Youth Action Committee, held a public reception for the couple at the American Institute of Aeronautics and Astronautics in Los Angeles. Linus and Ava Helen also opened their home to celebrate with friends and former students. The celebrants made banners to honor the Paulings, one reading, "We knew you when you only had one." Another took on a more mathematical form: "LP plus AHP equals PAX plus 2 Nobels."

While the Paulings attended multiple celebrations in their honor, they were obliged to refuse other offers as they prepared for their trip. In one instance, a Caltech student requested that Pauling give a farewell speech before leaving for Scandinavia. Turning down the request, Pauling would tell the *Associated Press* a few days later that he was not giving any speeches before he accepted his Nobel Prize so as not to "be tempted to let something drop beforehand."

In this, Pauling may have been alluding to an increasingly untenable situation at his base institution. Much has been made of Caltech's official non-response to Pauling's Nobel Peace award. Indeed, the only recognition that occurred at all on the Pasadena campus was a small coffee hour hosted by the Biology Department. The fact that his own department, to say nothing of the larger institution, chose to ignore this major decoration was deeply hurtful to Pauling and, by December, he had already announced his departure from Caltech, his academic home of some 41 years.

For their part, mainstream journalists had finally begun to take a more amicable and celebratory approach in their portrayals of Pauling than had generally been the case since the announcement of his Nobel win in October. These articles provided a more personal reflection of Pauling while not completely ignoring the earlier controversy. Of particular note, as the Paulings flew across the Atlantic on December 7th, the *Honolulu Star-Bulletin* published an interview with their eldest son, Linus Pauling Jr., in which he spoke of his father's activism. When asked to gauge Pauling's talents, Linus Jr. downplayed his father's peace efforts in relation to his work as a scientist. "In terms of his ability to marshal facts and to organize and reorganize them toward a goal," the younger Pauling is quoted, "his strivings for peace are comparatively simple, whereas this ability applied to science demonstrates an astoundingly high degree of creative imagination. Essentially, his work in peace is a public relations job — to get the facts across to the public in a meaningful way."

When asked what inspired his father to engage in peace work, Pauling Jr. said that "innately, he has always been a humanitarian." His aversion to hunting and his resistance to Japanese internment during World War II were submitted as past evidence of these humanist proclivities. Digging a little deeper, Pauling Jr. then pointed to his father's experience with nephritis in the early 1940s as a turning point in taking on social issues. At the time of his diagnosis, the elder Pauling "was pretty close to dying." During his recovery he had no choice but to rest, an experience that "forced him to contemplate on the value of life; on the wrongs of killing and harming that was going on around the world." Ava Helen also was a major influence on her husband, according to her son. Of particular importance was her work in the early war years with Union Now, an activist platform that called for world government as a tool for mediating international issues between nations.

Pauling Jr. was also, and inevitably, asked to address some of the controversy surrounding his father's alleged Communist ties, and responded that he had "never heard him praise Communism, although, on occasion, he has criticized some aspects of capitalism, such as unequal opportunities." "Today," he continued, "that's known as civil rights."

The following day, an *Associated Press* article that centered on Pauling's life and his "book-and-paper strewn den" came out in newspapers across the country. With the Nobel ceremonies only two days away, the headlines that ran with the article insinuated something big, with variations of "Pauling Hints at Surprises in Oslo Speech," "Linus Pauling Promises Speech Shocker," or "Pauling Predicts Shock." The actual feature article, written by Ralph Dighton, was far less sensational.

A *Peanuts* cartoon strip signed by Charles Schultz on display at the Pauling home helped Dighton to characterize his subject. The strip showed the animated character Linus stacking blocks in a "gravity-defying stairstep fashion" with the tagline "Linus, you can't do that!" Pauling's own reaction after first encountering it, Dighton noted, was a loud laugh. For Dighton, the lesson of the cartoon was that "the world has been telling Pauling [you can't do] that all his life, and he keeps doing what for many others would be impossible."

In his piece, Dighton also spoke of Pauling's decision to leave Caltech in favor of the Center for the Study of Democratic Institutions. Pauling avoided any discussion of problems at Caltech and instead explained that receiving the Nobel Peace Prize was an opportunity, and his decision to leave was a direct consequence of his Prize. At the Center,

Pauling would have more freedom to focus on his peace work and international affairs while also continuing to delve into the molecular basis of disease.

While the article avoided adding to the controversy surrounding Pauling, it still discussed past instances to remind readers that Pauling was never far from trouble. The list was familiar to those who had followed Pauling's career: the 1952 denial of Pauling's passport due to suspected Communist proclivities; his 1958 petition against nuclear testing, signed by over 11,000 scientists from 50 different countries; the consequent 1960 "joust" with the Senate Internal Security Subcommittee; his picketing of the Kennedy White House just before attending dinner inside. All helped to exemplify how Pauling had "made a public scourge of himself."

*

As Pauling set his sights on Europe and the spotlight of international attention, he left behind a busy office. His assistants Helen Gilrane and Katherine Cassady, who had typed up numerous thank-you letters and helped to coordinate Pauling's increasingly busy life over the previous two months, continued to provide support from a distance. Both kept on answering Pauling's unceasing correspondence while also working out the details of his still-unfolding trip. For her part, Gilrane responded to Bertrand Russell among many others, telling him that Pauling would be unable to respond right away. Communications of high importance were forwarded to Pauling in Norway; others had to wait until he came back in January.

The Paulings' return trip through the East Coast had not been finalized either. Gilrane made reservations in New York and Philadelphia while also coordinating Pauling's wardrobe for the celebrations that would take place after his European journey. For one occasion she wrote to Samuel Rubin, President of the American-Israel Cultural Foundation, that Pauling "shall have his smoking jacket as well as his tails, but he will wear whatever you think is appropriate for the evening."

Pauling also benefited from the help of friends in Norway. Otto Bastiansen, a Norwegian physicist and chemist, worked out much of Pauling's Scandinavian itinerary, deciding where he would lecture and with whom he would meet. Since the entire Pauling family, including the children's spouses, was coming, Bastiansen had to work out how they might be involved in activities as well.

The Paulings' first event happened almost as soon as they landed in Oslo on Sunday, December 8th. Ignored by the U.S. embassy or any other official representative from his home country, Pauling was welcomed at a private party held at the home of Marie Lous-Mohr, a Norwegian Holocaust survivor and peace activist who had also helped to arrange Pauling's schedule. On December 9th, the day before the Peace Prize ceremonies, the Paulings had lunch with friends and colleagues from Norway at the Hotel Continental, where they were staying. Afterward Linus and Ava Helen, together with two of their sons, Peter and Linus Jr., participated in a press conference. The event focused mostly on the man of the hour, who was very optimistic about the significance of receiving the Nobel Peace Prize and expressed gratitude toward the many people who worked in the peace movement. The *Associated Press* quoted Pauling as stating,

> "I think awarding me the prize will mean great encouragement to the peace workers everywhere, but particularly in the United States, where there have been so many attacks upon the peace workers... For some time it has been regarded as improper there to talk about these things. I think this is about the best thing that could happen for the movement."

As the world was still mourning the assassination of President John F. Kennedy, Pauling said that the slain leader's "attitude" helped make peace activism acceptable in the United States. Pauling steered away from any criticisms of Kennedy, as one reporter prodded Pauling about his opinion of the 1962 Cuban Missile Crisis. In the midst of the crisis,

Pauling had been extremely critical, calling Kennedy's threat of military action against Russia "horrifying" as it could easily lead to the use of nuclear weapons. In Oslo, Pauling reframed his rhetoric, telling the press that the situation "taught the lesson to the world, that the existence of nuclear weapons is a peril to the human race."

The next day, Pauling would receive the Nobel Peace Prize for helping alert the world to that peril.

*

On December 10, 1963, Linus Pauling accepted the belated Nobel Peace Prize for 1962. Attended by the Norwegian royal family and various government representatives, the ceremonies took place in Festival Hall at the University of Oslo in Norway — separate, as per tradition, from the other Nobel Prize ceremonies in Stockholm, Sweden. Pauling shared the ceremony with the winners of the 1963 Nobel Peace Prize — an award split between the International Committee of the Red Cross and the League of Red Cross Societies, in commemoration of the centennial of the founding of the Red Cross.

Gunnar Jahn, the chair of the Norwegian Nobel Committee, introduced Pauling before presenting him with the Prize. In his remarks, Jahn reconstructed the advances and setbacks of the post-war peace movement in which Pauling had so prominently operated. In the months and years following Japan's surrender, escalating Cold War tensions had rendered as unlikely any hopes for an immediate era of peace. The nascent post-war peace movement, according to Jahn, "lost impetus and faded away. But Linus Pauling marched on: for him retreat was impossible."

While Pauling's peace work was surely political in nature, Jahn drew particular attention to the importance of Pauling's scientific attitude in researching the impact that atmospheric radiation could make on future generations. Even critics of Pauling, including Edward Teller, did not fundamentally disagree with him concerning the harmfulness of fallout from nuclear tests. Where Teller and Pauling did conflict centered more on questions of whether or not these harmful effects outweighed the advantages that they provided to the United States with respect to the Soviets. Pauling thought they did; Teller disagreed.

Jahn then recounted how the public started paying close attention to Pauling in 1958 as he presented to United Nations Secretary General Dag Hammarskjöld a petition signed by 11,021 scientists from 50 different countries calling for the end of above-ground nuclear weapons testing. Because of the petition, Pauling was called before Congress and questioned about alleged Communist ties which, not for the first time, he denied. By Jahn's estimation, the hearings only served to make Pauling a more popular and sympathetic character, and he continued to speak out more and more.

For Jahn, Pauling's 1961 visit to Moscow, during which he delivered a lecture on disarmament to the Soviet Academy of Science, illustrated his importance in propelling the Partial Test Ban Treaty, which came into effect two years later. While there, Pauling unsuccessfully sought to meet with Soviet Premier Nikita Khrushchev. Unbowed, he instead sent Khrushchev two letters and a draft nuclear test ban agreement. "In the main," Jahn emphasized, "Pauling's proposal tallies with the test-ban agreement of July 23, 1963. Yet no one would suggest that the nuclear-test ban in itself is the work of Linus Pauling... But, does anyone believe that this treaty would have been reached now, if there had been no responsible scientist who, tirelessly, unflinchingly, year in year out, had impressed on the authorities and on the general public the real menace of nuclear tests?"

Ultimately it was as a scientist that Pauling helped move the world toward peace. Looking forward, Pauling's proposed World Council for Peace Research would bring together bright minds from the sciences and humanities under the umbrella of the United Nations to seek out new institutional models and paths of diplomacy for a nuclear-armed world. Jahn closed by suggesting that "through his campaign, Linus Pauling manifests the ethical responsibility which science — in his opinion — bears for the fate of mankind, today and in the future."

After concluding, Jahn called Pauling to the stage; applause and a standing ovation from the crowd quickly followed. After the cheering had died down and Jahn presented Pauling with the gold Nobel medal and a certificate, Pauling delivered a brief acceptance speech, calling the Prize "the greatest honor that any person can be given." But Pauling also recognized that his Prize was a testament to "the work of many other people who have striven to bring hope for permanent peace to a world that now contains nuclear weapons that might destroy our civilization."

Pauling went on to draw similarities between himself, the first scientist to be awarded the Nobel Peace Prize, and Alfred Nobel, who endowed the Nobel Foundation. Both were chemical engineers interested in scientific nomenclature and atomic structure. Both owned patents on explosive devices — Nobel the inventor of dynamite and Pauling an expert on rocket propellants and explosive powders whose skills came to bear during World War II. And both expanded their interests into biology and medicine as well. Many had described Nobel as a pessimist, but Pauling assured his audience that this was not the case. Rather, like himself, Nobel was an optimist who saw it as "worthwhile to encourage work for fraternity among nations."

The following day, December 11th, Pauling gave his Nobel Lecture, "Science and Peace." In it he described how the advent of nuclear bombs was "forcing us to move into a new period in the history of the world, a period of peace and reason." Development of nuclear weapons showed how science and peace were closely related. Not only were scientists involved in the creation of nuclear weapons, they had also been a leading voice in the peace movement, bringing public awareness to the dangers of such weapons.

Pauling recounted how Leo Szilard — whose 1939 letter to President Roosevelt (co-signed by Albert Einstein) had led to the Manhattan Project — urged Roosevelt in 1945 to control nuclear weapons through an international system, a plea that was issued before the first bombs had been dropped. While Szilard's appeal fell flat, it was followed, in 1946, by the creation of the Emergency Committee of Atomic Scientists, a group overseen by Szilard, Einstein and seven others, including Pauling. Over the next five years, the committee warned of the potentially catastrophic consequences of nuclear war and advocated for the only defense possible: "law and order" combined with a "future thinking that must prevent wars."

Other groups followed. For Pauling, the Pugwash Conferences, headed by Bertrand Russell from 1957 to 1963, were particularly influential in bringing attention to the harmful effects of nuclear testing and, ultimately, the 1963 Partial Test Ban Treaty. It was during this time that nuclear fallout, the subject of Pauling's 1958 petition, became of greater concern. The importance of fallout centered on the potential genetic mutations to which several generations would be exposed. Pauling quoted the recently assassinated President Kennedy to support his point: "The loss of even one human life, or malformation of even one baby — who may be born long after we are gone — should be of concern to us all."

As matters stood in 1963, Pauling feared that the time to effectively control nuclear weapons was fast slipping away. The test ban treaty was, Pauling lamented, already two years too late and had not prevented the large volume of testing that had taken place after the Soviets — who were quickly followed by the United States — had broken the 1959 testing moratorium in 1961. The failure to end testing outright before 1960 had led to the explosion of 450 of the 600 megatons detonated during all nuclear tests.

Because of the sheer number of nuclear weapons in existence (Pauling estimated some 320,000 megatons), limited war was no longer feasible, due to "the likelihood that a little war would grow into a world catastrophe." Abolishing all war was the only way out. But standing in the path of the abolition of war were people in powerful positions who did not recognize or acknowledge the present dangers. Pauling also saw China's exclusion from the United Nations, which prevented the nation from taking part in any discussions on disarmament, as an additional roadblock to a lasting world peace.

To get around these blockades, Pauling proposed joint national and international control of nuclear weapons as well as an inspection treaty aiming to prevent the development of biological and chemical weapons. Additionally, Pauling felt that small-scale wars should be abolished and international laws established to prevent larger nations from dominating smaller ones.

The challenges of the era were great but Pauling ended optimistically:

> "We, you and I, are privileged to be alive during this extraordinary age, this unique epoch in the history of the world, the epoch of demarcation between the past millennia of war and suffering and the future, the great future of peace, justice, morality and human well-being... I am confident that we shall succeed in this great task; that the world community will thereby be freed not only from the suffering caused by war but also, through the better use of the earth's resources, of the discoveries of scientists, and of the efforts of mankind, from hunger, disease, illiteracy, and fear; and that we shall in the course of time be enabled to build a world characterized by economic, political, and social justice for all human beings, and a culture worthy of man's intelligence."

The response to Pauling's speech by the American press was fairly tame. Most headlines simply issued variations on "Pauling Gets His Prize." A handful of headlines delved into the substance of Pauling's lecture, one noting "Pauling Accepts Award, Sees World without War in Sight." Others emphasized the means by which he sought to end war, e.g., "Pauling Urges UN Veto Power on Nuclear Arms."

The substance of the articles, most of which relied upon *Associated Press* copy, continued to focus on Pauling's past controversies and suspected Communism. From his lecture, the reports tended to highlight his homage to the late President Kennedy and the dollar amount of his Prize. When Pauling's policy proposals came up, mostly in larger papers that did not rely on the *Associated Press*, China's admission to the United

Nations and UN veto power over the use of nuclear weapons were seen as relevant and potentially controversial.

Absent from the press coverage was any discussion of the scientific components of Pauling's lecture. This included his claims concerning the harmful health effects of nuclear weapons as well as his descriptions of the increases in size and number of nuclear weapons.

One bit of critical commentary, published in the *Wall Street Journal*, came out a week after Pauling's speech. In it, author William Henry Chamberlin dismissed Pauling's views on peace as both unpopular and overly simplistic. Pauling's reasoning ran counter to the thinking of all US presidents since Truman — namely, that the only avenue to peace is to make as many weapons as the Soviets. Chamberlin noted that even scientists — specifically Edward Teller — agreed.

In Chamberlin's estimation, Pauling was fundamentally an alarmist. Further, Pauling had no impact whatsoever on the Partial Test Ban Treaty. The idea for the treaty had emerged out of the governments of the United States and Great Britain long ago, and its delay in ratification was due solely to foot-dragging from the Soviets. Chamberlin also discounted Pauling's claim to be a representative of a worldwide movement for peace by characterizing his efforts as "a one-man crusade."

On the same day that Chamberlin's piece was published, the Portland *Oregonian* released a short article of its own, profiling Pauling's freshman physiography teacher at Washington High School, Pauline Geballe. Pauling pointed to her as one who had helped to ignite his interest in science, and the two had remained in touch over the years. Geballe herself, through the League of Women Voters, was also part of the peace movement. On behalf of the group, she had recently queried Pauling for insight into questions of disarmament. Pauling responded by sending her a copy of his book, *No More War!*, from which Geballe read aloud the next time the group met. Geballe and her colleagues seemed to evidence that, just as he had been stating for the previous two months, and likewise in his Nobel acceptance speech, Pauling was merely a representative of a much larger movement, if still a polarizing and extremely prominent one.

*

The formal ceremonies completed, Linus and Ava Helen remained in Scandinavia into the New Year, visiting friends and making several public appearances in the region. In the wake of his Prize, Pauling continued to speak on the importance of peaceful international relations and also addressed scientific audiences. Often, the two would overlap.

For the rest of their first week abroad, the Paulings remained in Oslo, where the whole family was able to spend some time together and mingle informally with others. On Friday, December 13th, the Paulings visited the Munch Museum, which had just opened to mark the 100th anniversary of the birth of Norwegian painter Edvard Munch. Afterwards, they attended a luncheon hosted by the Norwegian Chemical Society. The weekend continued with informal meetings and a party at the home of Otto Bastiansen.

The following Monday, the Paulings flew north to Trondheim where Linus gave a Normann Lecture, part of a popular series on philosophy and science sponsored by the estate of local fishmonger Evard Normann. Pauling's talk, "Humanism and Peace," continued in the vein of his Nobel Lecture by addressing the need for humanity to decide between peace reached through reason versus annihilation brought on by war. Pauling emphasized that people were no longer separated into small groups where communication was easier. "Now we are all bound together," he said. "One organism. Will it survive or become extinct?"

Pauling then referred to Aristotle's assertion that nations were inherently immoral, suggesting that "Now they will become moral." The next day, Pauling kept up his slate of public appearances by meeting with local high school students and giving a lecture to the Trondheim section of the Norwegian Chemical Society on the role of molecular diseases in evolution.

The Paulings spent the rest of the week in Sweden. Their first stop took them to Stockholm, where they were immediately met by a collection of prominent Swedish scientists and politicians. Later, Pauling addressed the press before attending a reception hosted by the local Chapter of the Women's International League for Peace and Freedom. The next morning, Linus gave another lecture on molecular disease and its relation to evolution, this time to the Swedish Royal Institute of Technology. In the afternoon, the Paulings visited the newly opened Wenner-Gren Center, which Hugo Theorell, head of the Nobel Medical Institute's Biochemistry department, described to Pauling as representing "one of the many ways which have to be tried in order to improve mutual understanding and friendship between different people and races." Devoted to scientific collaboration across borders and staffed by researchers from 30 countries, Wenner-Gren struck Pauling as a good example of how science could foster peaceful international cooperation.

That evening, local peace groups held a torchlight procession in Pauling's honor. The march ended in front of the Storkyrkan, the oldest church in Stockholm, where Pauling gave a speech on peace. The talk was similar to the Trondheim address, referring again to Aristotle's notion that nations are inherently immoral, since, as Pauling quoted, "It is considered proper for a strong nation to attack a weak one, if she can thereby benefit herself, irrespective of what the principles of morality have to say about this action."

Pauling also condemned the United States' rejection of the 1957 Rapacki Plan, which would have established a nuclear-free zone in central Europe. "Up to that time," Pauling offered, "everything had been simple. Everything could be blamed on the Russians." Afterward however, Western intellectuals began to lose faith in the notion that the "free world" truly sought "peace, democracy, and a better future."

From Stockholm, the Paulings went southwest to Gothenburg and then south again to Lund, stopping at both cities' universities to speak on Linus' own path to becoming a peace activist. From there the couple returned to Oslo and enjoyed a more relaxed schedule over the holidays.

Before heading back to the United States, the Paulings spent a day in Copenhagen, where Linus would give one final talk on peace. According to a letter to Pauling from Gerda Ottesen — wife of Martin Ottesen, the director of the Carlsberg Laboratory — Pauling's time there had inspired the formation of a "Pauling Group" that endeavored to translate his speeches and writings on peace into Danish. Ottesen also informed Pauling that he had become of interest to a newly formed political party that had emerged out of disputes from within the Danish socialist party. His mark clearly made in Scandinavia, Pauling headed back to the States.

The Paulings arrived in New York City on Saturday, January 4th and kept a low profile during their first few days on home soil. On Wednesday they flew to Washington, D.C. for a press conference — their first public appearance in the U.S. since Pauling had been awarded the Nobel Prize. The Women's National Press Club hosted the press conference, according to president Elsie Carper, so that Pauling could talk about scientists' responsibilities in a democratic society, and to "promote better understanding" of science in "world affairs."

Despite the event's stated intention, what ended up in print contained little in the way of science. As with mainstream reporting from the Nobel ceremony, writing on the press conference focused mostly on Pauling's previous statements about the need for China to be involved in disarmament negotiations and his much published suggestion that "there will never again be a world war — or any war in which nuclear weapons are used — or any great war."

Other aspects of the press conference dug more into Pauling's political interests. During the question-and-answer portion, one reporter asked Pauling if he would ever run for office. Pauling said he would rather work on his science and for peace independently, and that he was too "selfish" to run for office. He spoke as well of growing up in a Republican family and noted that he did not become a Democrat until 1932, when so many others also changed their minds in the midst of the Great Depression. Pauling added, "I have been one ever since."

The media event concluded, the Paulings hurried back to New York City that afternoon so that they could attend a tribute held at the Hotel Commodore. Sponsors of the event included several scientists from around the world, as well as Bertrand Russell, Walter Cronkite, Senator George McGovern, and Arthur Miller. The gala was also attended by United Nations Secretary General, U Thant.

Pauling gave a short speech during the proceedings but was just one among many who spoke — Ava Helen, Dagmar Wilson of Women Strike for Peace, and historian Henry Steele Commager also enjoyed their moment at the podium. For his part, Pauling gave a chilling account of how many nuclear weapons were then in existence, stating, "If there were to take place tomorrow a 6-megaton war, equivalent to the Second World War in the power of explosives used, and another such war the following day, and so on, day after day, for 146 years, the present stockpile would then be exhausted."

The Paulings rested the next day and then took to the skies again, this time destined for Philadelphia, where they made two more public appearances. On Thursday, January 9th, local peace groups, including Women Strike for Peace and the Society for Social Responsibility in Science, sponsored another tribute. In addition to remarks from Pauling and others, those attending were availed of the opportunity to sign pre-printed postcards addressed to President Lyndon Johnson, applauding his call for a "Peace Offensive" while continuing to urge him to negotiate for complete disarmament. The following night, the local Women Strike for Peace Chapter held a cocktail party in honor of the Paulings.

The Paulings headed back to New York on Saturday where they would attend one more reception in their honor, this time hosted by the Women's International League for Peace and Freedom. On Wednesday they, at long last, flew back to California, thus concluding their five and a half weeks of travelling, lecturing, and celebrating in connection with

Linus Pauling's Nobel Peace Prize. Once home, Pauling did not rest for long: within days he was in public again giving speeches outlining what he saw as the "Next Steps" to reach disarmament and peace in the world.

15 William P. Murphy

By Ben Jaeger

William P. Murphy, 1935.

Linus Pauling's boyhood home of Condon, Oregon is a very small farming town located in dryland wheat country in the north-central portion of the state. According to the 2020 census, the town's population consisted of only 711 people, about the same number as when the Pauling family lived there in the early 1900s. Despite its size, Condon can boast of an interesting statistic: two of its early 1900s residents would later go on to win Nobel Prizes. Pauling, of course, is one; the other is a physician named William P. Murphy.

William Parry Murphy was born on February 6, 1892 in Stoughton, Wisconsin. He was educated at public schools throughout Wisconsin and Oregon and, in 1909, was a member (along with five others) of Gilliam County High School's first graduating class. Murphy moved to Condon when he was 15, meaning that his time there did overlap with Pauling's. It is unlikely that they would have had much in the way of contact with each other however, as Pauling was a full nine years younger than Murphy. (We do know, however, that the two men corresponded later in their lives and that Murphy was a signatory of Pauling's famous United Nations Bomb Test Petition.)

Following his high school graduation, Murphy attended the University of Oregon in Eugene. In 1914 he received an A.B. degree, and spent the next two years teaching physics and mathematics in high schools around the state. After his short stint as a teacher, Murphy decided to shift towards medicine. He spent one year at the UO Medical School in Portland, and appears to have left Oregon for good that summer when he enrolled in courses at the Rush Medical School in Chicago.

On September 10, 1919, Murphy married Pearl Harriet Adams, and that same year was awarded the William Stanislaus Murphy Fellowship at Harvard Medical School in Boston. (The name of this fellowship is important as, per the donor's decree, it was awarded only to individuals with the surname "Murphy," spelled exactly that way.) He retained this fellowship for three years, and graduated from Harvard in 1922 as a Doctor of Medicine. His schooling completed, Murphy then interned at the Rhode Island Hospital in Providence. Not long after, he was appointed resident physician at Peter Bent Brigham Hospital in Boston.

During his short time at the University of Oregon Medical School, Murphy had become interested in developing a cure for anemia. However, he wasn't able to actively pursue this ambition until he was at the Boston hospital. He specifically began to work on pernicious anemia, using intramuscular injections of liver extract as a treatment for both pernicious anemia and hypochromic anemia.

After his time at Brigham Hospital, Murphy returned to Harvard where he was appointed Instructor of Medicine. It was there that he collaborated with George Richards Minot and George Hoyt Whipple to develop a treatment of pernicious anemia through a diet of uncooked liver. In 1934, Murphy, Minot, and Whipple shared the Nobel Prize for Physiology or Medicine "for their discoveries concerning liver therapy in cases of anaemia." Five years later, Murphy compiled this work into a text titled *Anemia in Practice: Pernicious Anemia.*

Although Murphy never lived in Oregon again, he did occasionally return to visit his parents, sister, and two brothers, all of whom had relocated to Portland. On one occasion, not long after winning the Nobel Prize, Murphy was interviewed for an article in *The Oregonian* newspaper. When asked about his research, he noted:

"A few years ago pernicious anemia was one of the diseases that was always fatal, it was sure to be fatal within a year. But now it need not be so any more than measles or other minor complaints. A person who has pernicious anemia has a life expectancy as long as if he didn't have it, providing the proper treatment is given."

(Sadly, Murphy's breakthroughs were not made soon enough for Pauling's mother Belle, who died of pernicious anemia in 1926.)

Throughout the remainder of his career, Murphy worked as a consulting hematologist to several hospitals. He also made his way through the ranks at Harvard Medical School, moving from Assistant in Medicine in 1924 to Senior Associate in Medicine in 1958. After his retirement, he was appointed Emeritus Lecturer.

Although he shared the Nobel Prize, leaders in the nation of Finland called him the "real discoverer" of the cure for pernicious anemia and gave him the Order of the White, the country's highest decoration. He also received the Cameron prize from the faculty of the University of Edinburgh, the Gold Medal from the Massachusetts Humane Society, and was elected a member of the Halle Academy of Science in Germany. He passed away on October 9, 1987.

16 The *National Review* Lawsuit

By Jessica Newgard

National Review Sued for Million By Pauling Over Traitor Label

Dr. Linus C. Pauling, Nobel-prize winner and an advocate of nuclear disarmament, has filed a $1,000,000 libel suit against the conservative weekly, National Review, its publisher, William A. Rusher, and its editor, William F. Buckley Jr.

Dr. Pauling charges that the magazine, in articles published on last July 17 and Sept. 25, falsely branded him "a traitor, a collaborator with subversive foreign and alien elements . . . engaged in subversive Communist activities" and "a moral nihilist."

The suit became known yesterday when the scientist's lawyer, Michael Levi Matar, filed a pre-trial motion in State Supreme Court, seeking to vacate the defendants' demand for a bill of particulars.

The first National Review article mentioned Dr. Pauling along with Mr. and Mrs. Benjamin J. Buttenweiser, the late Prof. C. Wright Mills of Columbia, Prof. Frederick V. Schuman of Williams, Prof. Thomas I. Emerson and Prof. Fowler V. Harper of Yale. It

The New York Times
Dr. Linus C. Pauling

pleaded the defenses of truth,

Despite Pauling's dismissal by the Senate Internal Security Subcommittee as described in Chapter 11, numerous articles were published in newspapers and magazines around the country that decried Pauling as a Communist supporter and criticized his refusal to release the names of the people who had help to collect signatures for the bomb test petition. One of those articles, titled "Treason à la Mode," was published by

National Review magazine on December 31, 1960. In it, author James
Burnham wrote,

> "Linus Pauling is still at large and unindicted for his contemptuous
> refusal to give the Internal Security Subcommittee the facts about
> how thousands of names of scientists — including several thousand
> from Communist nations — were collected for his petition against
> H-bombs. While Pauling propagandizes for policies which corrode
> American morale and promote the interests of the Communist enter-
> prise, he continues to enjoy wide popular esteem, security in his pro-
> fessorial post at Caltech and large audiences on the university circuit...
> The point is not whether these men are conscious traitors, which in
> all, or almost all, cases they are not. But the Communists are traitors.
> These men, by their acts, have condoned the Communist enterprise
> and advanced its interests. Our society, by condoning the actions of
> these men, condones also the enterprise."

A full year and a half later, *National Review* published a second
article critical of Pauling, this time an editorial titled "The Collaborators," which went to print on July 17, 1962 and again raised the specter
of treason.

> "Professor Linus Pauling of the California Institute of Technology, [is]
> once more acting as megaphone for Soviet policy by touting the World
> Peace Conference that the Communists have called for this summer
> in Moscow, just as year after year since time immemorial he has given
> his name, energy, voice and pen to one after another Soviet-serving
> enterprise."

Pauling, unlike many Americans at the time, did not see the Soviet
Union purely as an enemy and was not afraid of Communism or its perceived consequences. On the contrary, he believed that the best way to
ensure world peace and to promote the advancement of science was
to form mutually beneficial partnerships between communities. In this
vein, Pauling maintained a cordial relationship with many Russians and

traveled to the USSR six times, mainly to talk about science, but also to promote an end to nuclear bomb testing. His inviting stance towards the Soviet Union was seen by some Americans as pro-Communist and anti-American, but Pauling never identified as a Communist and was a strong believer in democracy.

The provocations from *National Review* came during a decidedly litigious period for Pauling. By 1962 he had already successfully sued the *Bellingham [Washington] Herald* for a defamatory letter to the editor published in December 1960. The case was settled for $16,000 and a retraction printed by the paper. Pauling also had three other court cases in motion during this period: libel complaints against Hearst Publishing Co. and King Features Syndicate for $1 million, the *St. Louis Globe-Democrat* for $300,000 and the *New York Daily News* for $500,000.

Pauling worried that the deluge of articles attacking his reputation and labeling him a Communist would decrease sales of his textbooks and negatively affect his position at Caltech. He later testified that libelous statements had cost him a raise from Caltech in 1962 and had resulted in his being treated coldly by the Caltech president and others on campus. He also confided that his book income had gone down slightly and that he had suffered a loss of self-confidence.

Determined to restore his good name, Pauling contacted the lawyer whom he had enlisted for the Hearst Corporation lawsuit, Michael Levi Matar, about the *National Review* articles. Matar replied that the pieces had libeled Pauling "as a Communist and moral nihilist...[and there is] little doubt in your case that malice by these defendants would not be too difficult to establish." The duo decided to sue the magazine for $1 million.

*

Upon learning about Pauling's lawsuit, *National Review* went on the offensive, publishing an article in September 1962 titled "Are You Being Sued by Linus Pauling?" As they answered their own question, the magazine's publishers pulled few punches in reaffirming the stance that had so angered Pauling in the first place:

"We are (or so his lawyers tell us). And so are other well-behaved papers
and people throughout the country. Professor Pauling...seems to be
spending his time equally between pressing for a collaborationist foreign policy, and assailing those who oppose his views and who question
whether this country can simultaneously follow Dr. Pauling's recommendations and remain outside the Communist orbit. Dr. Pauling is
chasing after all kinds of people... His victory signal is the check or two
he has wrested from publishers — who may indeed have libeled him, in
which case they should pay up; but who may simply have been too pusil-
lanimous to fight back against what some will view as brazen attempts
at intimidation of the free press by one of the nation's fellow travelers."

This public response made it clear that the *National Review*'s
founder, William F. Buckley, had every intention of fighting the lawsuit
to its end.

At the time, Buckley's response probably seemed disingenuous to
Pauling, since he had already won a settlement against the Bellingham
paper and was in the midst of a complaint against the Hearst Corporation that would result in his receipt of $17,500 in June 1963. A twist was
soon to arise, however; one that made a major impact on the strategies
contemplated by Pauling and many others used to living their lives in
the spotlight.

New York Times v. Sullivan

Pauling officially filed suit against *National Review* on January 17, 1963,
but his case wound up being delayed for more than three years. This
lengthy pause had many causes, including the need for information gathering; Pauling's receipt of the Nobel Peace Prize in December 1963;
and Buckley's candidacy for mayor of New York City in 1965. The slowness of the proceedings also proved to be crucial as, in 1964, a landmark
Supreme Court case, *New York Times* v. Sullivan, was decided.

New York Times v. Sullivan is the United States Supreme Court
decision that confirmed the media's right to comment freely on public

officials without worry of being sued for libel, except in cases of provable "actual malice." As part of this decision, the Court established the definition of "actual malice" as a publisher knowing that a published statement was false or acting in reckless disregard of its truth or falsity at the time of printing.

From there, the burden of proof in a libel case would fall on the plaintiff, who was charged with establishing "actual malice" on the part of the defendant. Doing so is generally fairly difficult as there is often little evidence documenting the details of a defendant's motivations or intent. The Court handed down this ruling in order to protect the First Amendment and it proved to be hugely influential. Aside from cases like Pauling's, *New York Times* v. Sullivan made a significant impact on the Civil Rights Movement as, prior to the ruling, media in many southern states were wary of reporting on civil rights abuses for fear of libel action.

The *Times* decision did allow for future case-by-case determination of identifying parties subject to the burden of proving "actual malice." Importantly, the possibility of extending the *Times* ruling beyond public officials to private citizens seems to have been suggested first (or at least very early) by the United States Court of Appeals for the Second Circuit in Pauling v. News Syndicate Company, tried just four months after the *Times* case.

Decisions made in Pauling's lawsuit against the News Syndicate Company helped to establish a precedent of extending the *New York Times* v. Sullivan ruling from public *officials* to non-elected public *figures*. If his scientific work hadn't already done the trick, by 1964 Linus Pauling had certainly made himself a public figure through his high-profile peace activism, to say nothing of two Nobel Prizes. In determining that the *Times* ruling applied to him as well, the courts began to expand the restricted definition of libel to parties outside of public officialdom, an evolution that proved very damaging to Pauling's litigious streak.

In March 1966 the *National Review* case finally went to trial in the New York State Supreme Court, a point by which Pauling had lost essentially all of his legal traction. By then, the courts had already begun to apply the *Times* ruling to public figures, and Pauling was unable to prove the new definition of "actual malice" against the *National Review*. As such, it came as no surprise when, in April, the New York court held that

> "The constitutional guarantees [of the First and Fourteenth Amendments] require, we think, a federal rule that prohibits a public official from recovering damages for a defamatory falsehood relating to his official conduct unless he proves that the statement was made with 'actual malice' — that is, with knowledge that it was false or not."

Although Pauling was not a public official, the incident

> "would seem to favor extending the doctrine of that case at least to a private person who 'has thrust himself into the vortex of the discussion of a question of pressing public concern.' ...It is clear that if any private citizen has, by his conduct, made himself a public figure engaged voluntarily in public discussion of matters of grave public concern and controversy, Dr. Pauling has done so. Finally, the criticisms made of him in the alleged libelous articles are not criticisms of his private life; they are criticisms of his public conduct and of the motives for that public conduct."

Pauling understood that legal precedent was weighing against him but disagreed with the Supreme Court's stance on *New York Times v. Sullivan* — in effect, he felt that the tide of legal opinion weighing against him was itself fundamentally incorrect. Likewise, he objected to the court's definition of "actual malice," writing

> "I believe that the new criterion, as stated above, should not be upheld by the Supreme Court of the United States, because it places a reward on irresponsibility. I believe that the utterances of false and defamatory statements should be required by law to assume some responsibility, even though the false and defamatory statements are made about public officials or public figures."

Bruised by his *National Review* loss and disappointed by the *New York Times* verdict, Pauling actually contacted another lawyer, Louis Nizer, in hopes of finding a way to reverse the Supreme Court's thinking on the standard of libel for public figures. Nizer, while pessimistic, agreed to launch an appeal of the public figures extension of the *Times* case. This appeal was pursued for three years until the firm ultimately conceded to the dismissal of its case in 1968.

In the wake of the bitter *National Review* saga, Pauling ended up either losing or abandoning all of the other cases that he had in motion. His legal power had been taken away by *New York Times* v. Sullivan and, though often provoked over a lifetime that remained controversial, he never again mounted another libel case.

17 Peter Pauling

By Matt McConnell

(From left) Crellin, Linda and Peter Pauling, 1952.

Peter Jeffress Pauling was born on February 10, 1931. His middle name was given in honor of Lloyd Jeffress, his father's childhood friend, fellow undergraduate at Oregon Agricultural College, and best man on his wedding day (see Chapter 8). Peter was the second of four Pauling children. His older brother, Linus Pauling Jr., was born in 1925; Peter was joined swiftly by his younger sister Linda Helen in 1932 and, finally, by the baby of the bunch, Edward Crellin Pauling, in 1937.

In the early 1930s, everything seemed to be falling into place for the Pauling family. The same year that Peter was born, his father was promoted to full professor at the California Institute of Technology in

Pasadena, where the elder Pauling had been since 1922. The growing family moved into a new home on Arden Road shortly before Peter's birth, a transition that provided more space for the children and for their dog, Tyl Eulenspiegel, a cocker spaniel named after a German comic character. The year 1931 also marked Linus' first published article on the nature of the chemical bond — work that would ultimately result in a Nobel Prize.

In the decade following Peter's birth, his father was incredibly busy, giving 14 guest lectures at Berkeley alone before Peter turned three. He also compiled his revolutionary structural chemistry research in a groundbreaking book, 1939's *The Nature of the Chemical Bond*, a text authored while Pauling was in the midst of a series of 19 resident lectures at Cornell.

By the time of the book's publication, Pauling was becoming a major scientific figure, and as he began to travel more frequently, the children's mother, Ava Helen, would increasingly accompany him. During this period, the three youngest — especially Linda and Crellin — were often left in the care of a woman named Arletta Townsend, who became something of a second mother to them in their youth. The oldest child, Linus Jr., also shared in the supervision of his younger siblings from time to time, frequently trundling little Peter around the front yard in a child's red wagon, perhaps an early indicator of the boys' shared love of automobiles that would bring them together later in life.

The older kids also cared for the family rabbits, which their father raised at home for future use in his research. What was, for the children, a chore was also a reflection of Pauling's growing interests in serology, hemoglobin, and the formation of antibodies and their interaction with antigens. As his investigations moved forward in the late 1930s and early 1940s, Pauling found it either impractical or inconvenient to arrange for his study animals to be housed and tended at Caltech. He opted instead to build roughly 50 rabbit hutches on his own property. While the elder Pauling inoculated the rabbits and carefully recorded physiological data, his boys were charged with feeding, watering, and keeping the hutches clean.

At this time, Peter seemed to have little in common with his older brother, outside of their shared responsibilities around the house. Looking back, Linus Jr. would remember that most of their interactions revolved around fighting over the bathroom. An argument over this space once became so heated that it resulted in Linus Jr. splintering a door frame in the scrum.

The United States' entry into the Second World War touched the lives of all the Pauling children in different ways. In 1937, when Peter was six, an Army rifle range was built on the other side of a canyon near the Paulings' new home on Fairpoint Street. With this facility came a small group of Army guards, who would stand vigil near the family's property. Eventually, the soldiers became a social outlet for young Peter, who even as a child seemed to have a natural charm that often brought him rewards. For several years, Peter used the sentries as a means for procuring souvenirs and even stray military equipment. In the decades that followed, much of this treasure could still be found in storage at the family ranch near Big Sur, California.

Not only was Peter charming, he was quite intelligent as well. From 1936 to 1941, he attended Polytechnic Elementary, a private school in Pasadena. A 1937 report on his performance included the comment that he was, "Not only a superior child in intelligence, but one of the cutest children we ever took into the school." Two years later, the praise had only grown, with one educator writing that Peter "seems to be his Daddy's own boy, and that is saying a great deal."

From early on, Linus and Ava Helen harbored ambitious ideas for the pursuits that their kids should entertain, and Peter was identified as the child most likely to succeed in a career in science. Later in life, the elder Pauling would make passing insinuations about their potential as a father and son scientific duo, referencing the notable case of William and Lawrence Bragg, who shared the 1915 Nobel Prize in Physics for their work on X-ray crystallography. Peter would later become well-acquainted with Lawrence Bragg and his wife during a stint at Cambridge, a time period during which Peter had begun actively pursuing his own career in physics and chemistry.

In 1941, after Peter had completed his fifth year at Polytechnic, Linus and Ava Helen concluded that their gifted son had started to lag academically. To provide for a more effective learning environment, the Paulings had Peter — now ten years old — enrolled at Flintridge Preparatory School in Pasadena. Flintridge was a school for boys where Peter would live in a dormitory, his days closely supervised and specifically structured according to a daily schedule that ran from 7 A.M. to "lights out." At Flintridge, pupils were allotted one hour and 45 minutes of leisure time per day, with any other time not spent in class devoted to eating, completing chores, playing supervised sports or performing other physical exercises, or studying (also supervised). The school tailored its curriculum to the expectations put forth by a student's desired college or university, all with an eye toward ensuring that the student would be best prepared to pass entrance exams or to enter a university without taking the exams at all.

The school's approach was based on a three-fold training style that aimed to educate the mind, body and spirit, as evidenced by its motto, *Vires Corpore Mente Spiritique*. To chart the progress of the body, Peter's performances in a range of physical activities — from running and high jump to shot put — were recorded and graphed over the course of the year. These data were then plotted against an average representing the capacity of most boys his age, a practice referred to as a Physical Quotient plan. According to an informational pamphlet published by the school in 1941, Flintridge was the only school of its era that adhered to such a plan. The outcome, it promised, would be young men driven to develop their posture and muscle skills.

Once Peter was enrolled, his parents were issued monthly reports on their son's progress, and to their pleasure, his schoolwork showed measurable improvement. Unfortunately, the decision to move Peter to Flintridge caused Polytechnic to withdraw its scholarships for the remaining Pauling children. These scholarships had been provided with the contingency that the three eldest Pauling children attend the school, but with Linus Jr. completing his education at Polytechnic and Peter moving on to a different institution, it was no longer deemed appropriate to provide a scholarship for Linda alone. As a result, the family's only daughter was no longer able to attend private school at Polytechnic.

For Linus and Ava Helen, the summer months following Peter's first year at Flintridge were mostly spent abroad. As a result, the three youngest children were sent off to Camp Arcadia at Big Pines Park, California, while Linus Jr., now 16, was allowed to remain at home. While at camp, Linda became ill and was sent back a month early, leaving Peter and Crellin — from their perspective — stranded at the camp and feeling homesick. Linus Jr. wrote to Peter during this period, urging him to appreciate the time away from the mundane concerns of the Pauling home, including the return of parental discipline. "They've found all the things I didn't do and should have done, and all the things I did do and shouldn't have done," he wryly confessed.

In 1943, when Peter was 12, Linus and Ava Helen received letters from his teachers at Flintridge warning that his progress seemed increasingly hampered by an inability to focus on his work. The staff believed that Peter, like many intelligent children, was not being challenged enough academically, and that in his boredom he preferred to spend his time socializing rather than studying.

Perhaps only coincidentally, this was the same year that Peter's older brother turned 18 and left home for Berkeley. While Peter was being groomed in hopes that he might emerge as the next genius of the family, his older brother, Linus Jr., deliberately shied away from any such prospect as he entered adulthood, eventually postponing college aspirations and joining the Army Air Corps during World War II.

*

In April 1945, while German forces were surrendering to the Allies in Europe, Peter was completing his education at Flintridge Prep and moving on to McKinley Junior High, where he would enter the tenth grade. He continued to do well in most subjects, with the exception of a few poor marks in Latin. Now 14 years of age, Peter went outside the family home in Pasadena one day to discover a message painted on their garage door; it read: "AMERICANS DIE BUT WE LOVE JAPS. JAPS WORK HERE, PAULING." Peter quickly called for his parents, who surmised that the hate message had been written by misguided individuals angered by Ava Helen's work with the American Civil Liberties Union to prevent the internment of many Japanese-American citizens during the war.

Within the year, Linus Jr., now 21 years old, had returned home from his time in the Army Air Corps. He promptly came into possession of a 1932 Ford V-8 roadster that had belonged to a Mt. Wilson astronomer, Ted Dunham Jr. The car would become a marker of burgeoning adulthood for the Pauling boys, passing to Peter when Linus Jr. went off to medical school, and then again to Crellin when Peter finished college in California and left for Cambridge.

In 1947 Linus and Ava Helen returned from a scientific congress in Scandinavia to find their three youngest children growing somewhat depressed by, and resentful of, their frequent long absences. Knowing that they were about to spend six months in the U.K., where Linus would lecture as a visiting professor at Oxford, the Paulings decided it best to take the entire family abroad. In December they traveled by train to New York City, where they then boarded The Queen Mary and crossed the Atlantic.

The voyage would prove to be a huge missed opportunity for Linus. Onboard was Erwin Chagraff, who was excited to talk with Pauling about his discovery that DNA nucleotide base pairs obeyed a set rule — a 1:1 ratio of adenine to thymine and cytosine to guanine. As Crellin Pauling later recounted, "Chargaff had a reputation as a, well how do you put it politely, as a difficult personality. And what Daddy said to me was that he found Chargaff so unpleasant to be trapped on the Queen Mary with, that he dismissed his work." In doing so, Pauling overlooked the importance of a critical piece of information that would help lead Watson and

Crick to their discovery of the structure of DNA — a discovery with which both Linus and Peter would be intimately involved.

After the family returned from the U.K., Peter made the fateful decision to follow in his father's footsteps, enrolling at Caltech as an undergraduate and assuring his dad that his "chief purpose in life" was to be a physicist. Unlike the elder Pauling however, Peter gravitated almost immediately to those new freedoms that a young man no longer under his parents' roof might be expected to find inescapably important: cars, women, and parties.

With respect to the former, Peter wrote to his father from Caltech, asking whether or not Linus might be interested in helping to pay for a new engine for the roadster. Peter had already been putting some work into the car since it had passed from Linus Jr.'s hands into his own. Now off at college and free to pursue his own interests, he was eager to get under the hood. Peter likewise wrote to his mother asking for advice on different perfumes, listing the names of four different women, all of whom were apparently familiar to Ava Helen, and asking which scents she thought that each would prefer (he then added a fifth young lady to the list as an afterthought).

While Peter was at Caltech, the Pasadena house became something of a social hotspot for young, aspiring scientists, many of them graduate students or post-docs who coveted the opportunity to hobnob with the great Linus Pauling. By right of birth and strength of personality, Peter emerged as both gatekeeper and VIP at such events, and he thrived in this atmosphere. In his biography of Pauling's life, *Force of Nature*, author Thomas Hager paints a scene of pilgrims making their way up into the hills on warm afternoons for, "a beer, a dip in the pool, some jokes with Peter, and a chance to flirt with tall, slim, blond, teenaged Linda Pauling."

Peter, now 19 and cruising the streets of Pasadena at night in the modified "hand me down" roadster, was the life of many of these parties. His undergraduate years were accordingly marked by the ecstasies, despairs, and calamities — including a long and somewhat severe case of infectious mononucleosis — familiar to many college students. His father, observing from the middle distance, sent him a stern letter during this period, noting that he had opened some mail at the house intended for his son and that it was from the Bank of America, alerting Peter that his account was overdrawn by 50 cents. Pauling then advised his son, in great detail, about how best to manage his finances to avoid such a problem in the future.

Though they lived in the same city and worked at the same institution, Peter often corresponded through the mail with his father, expressing relatively little concern about his finances and far more with the prospect of being called up for the draft. In June 1950, North Korea, aided by the Soviet Union, invaded South Korea, where U.S. troops had been stationed since the close of the Second World War. As hostilities on the Korean peninsula ramped up, Peter grew increasingly fearful of being called upon to serve.

The elder Pauling advised that his son request a deferment as a student of physics, which Peter sought and successfully attained. Part of Peter's later motivation to spend some of his time as an undergraduate in London likewise emerged from his desire to avoid the draft for as long as possible. Peter felt that his enrollment as a student overseas would at least prolong his recruitment, whereas, if he remained at Caltech, he might he pulled in any day and waste his final undergraduate months on military training.

Meanwhile, in the summer of 1951, Peter was involved in a serious car accident. The *Pasadena Star News* reported that, "While driving his father's expensive 1949 sedan, Peter J. Pauling, 20, son of Linus Pauling, world-famed Caltech physicist, was injured in a spectacular traffic crash at Fair Oaks Avenue and Washington street." The police report indicated that the Pauling car had been sideswiped by Harry L. Nottingham, a 30-year-old welder, at 2:11 A.M.

The police jailed Nottingham overnight on a drunk-driving charge, and Peter was treated in the emergency room for mild injuries to his head that had been sustained when his car flipped onto its roof after the impact. The accident happened less than a month before Peter was to leave California to spend his summer at a laboratory in Woods Hole, Massachusetts. His eventual departure back east left his parents quite literally picking up the pieces in his absence, as Peter had requested that they keep what remained of the vehicle to see what he could salvage.

His father later wrote to Peter that, even with compensation paid out by insurance and the drunk driver involved, the family would still, after legal fees, "come out a little bit in the hole from your use of the Lincoln that night." Peter responded through his mother, writing, "Please tell Daddy that I am sorry I ruined his car," and asking that she remind him that, of the cars he could have wrecked, at least he chose the one that offered a barrier between his head and the road. The old Ford roadster was, after all, a convertible.

Peter's stint at Woods Hole was both formative and crucial to his next steps. While there, he used sodium and potassium tracers in squid axons to study ion movement in nerves. At the same time, he began seriously considering which institution to attend for graduate school, looking at Cambridge among others. Fortuitously, John Kendrew of Cambridge's Cavendish Lab was serving as a lecturer at Woods Hole and, unbeknownst to Peter, was recruiting for his protein structure research group.

Years later, Peter would recount that when Kendrew told Peter's boss at Woods Hole, David Nachmanson, that he had recruited him, Nachmanson replied, "What? That sex maniac?" Kendrew reportedly replied, "What does that matter?" In typical good humor, Peter offered that Kendrew had responded as such because he knew that being a "sex maniac" was an advantage at Cambridge. He later confessed, however, that his reputation was not well-earned, admitting that he was likely "the most unsuccessful Don Juan in Woods Hole."

While in Massachusetts, Peter was, in fact, pretty clearly agonizing over how to resolve his relationship with his college girlfriend, who remained in Pasadena. In his correspondence, Peter sometimes indicates a deep affection for his sweetheart, and at other times reveals significant doubt about any chance of a shared future. It is entirely possible that the hot and cold nature of Peter's feelings towards this woman were merely a reflection of a young man's whimsies. It might also be argued that this was an early sign of a lifelong struggle with manic depression that would come to plague Peter by the later stages of his graduate years in Cambridge. In any case, Peter's beau was equally unsure. At times she seemed to favor the appraisal of her father, a high-ranking scientific adviser to the American military who was stationed in Europe. The father believed that marriage would prematurely end his daughter's own academic ambitions and that, more broadly, Peter was bad news.

But by the following summer of 1952 — just before Peter left for the U.K. — she had warmed to her boyfriend again. It was a summer of exploration; the two crossed the nation prior to the beginning of graduate school for Peter, travelling together to New York, Washington D.C., Princeton, and Long Island. Their romance seemed to burn brightly, if briefly, as Peter's life in America drew to a close.

In the last few months before leaving the States, Peter and Crellin visited Hawaii, staying with their older brother Linus Jr., who lived there. Meanwhile, Linus and Ava Helen were engaged in world travels of their own, making lengthy stops in France and the U.K. Peter wondered aloud if he would get the chance to see his parents upon his brief return to Pasadena, or if, instead, he would be gathering his belongings from an empty house, departing with the well wishes of Linda and Crellin, and setting out alone for Montreal, where he would board a ship to cross the Atlantic. The exact circumstances of Peter's *bon voyage* from southern California are unknown, but by September 1952, he was on his way to a new life in the U.K.

*

"This tub moves steadily but slowly along." So wrote Peter Pauling in a letter to his mother, while riding somewhere in the Atlantic in the hull of a cargo ship that had been built in 1926. "It took us two and a half days to reach the open sea."

Having said goodbye to the nightlife of Montreal, and having entrusted brother Crellin with the needle to his old turntable, Peter took to the sea without much to his name save a bottle of duty-free Canadian Rye Whiskey; which, he lamented, did not keep him as warm onboard the ship as a good overcoat might have done. (Ava Helen would see to it that he had money to pick up some warmer clothes once he arrived in Cambridge, paid for with matured war bonds.) Onboard the ship, Peter shared his cramped cabin space with three roommates: a Scot, a "very pleasant and hard-working" Englishman, and an 18-year-old "pipsqueak" just out of rugby. Ever the charismatic socialite, Peter must have been excited to spend his days at sea with such an assortment of characters.

Not long after arriving in the U.K. in the fall of 1952, Peter began his studies at Cambridge University, working under John Kendrew, a Peterhouse Fellow in Max Perutz' Molecular Biology Unit at the Cavendish laboratory for physics. Although the Cavendish traditionally had not extended its focus beyond physics and physical chemistry to questions of biology, Sir Lawrence Bragg — director of the Cavendish and chair of the university's physics department — had recently supported an expansion of the lab's scope to include the mapping of biological structures.

This new Molecular Biology Unit would spearhead several important discoveries, among them Kendrew and Perutz's work on the atomic structure of proteins, the program of research that Peter was brought on to support and an accomplishment significant enough to garner the 1962 Nobel Prize in Chemistry. That same year, two other former Cavendish researchers — James Watson and Francis Crick — would receive their shared Nobel Prize in Physiology or Medicine for their discovery of the double helical structure of DNA, a breakthrough that Peter certainly observed from a front-row seat and even, perhaps, helped to make possible.

When Peter first moved into the office that he shared with Watson, Crick, and Jerry Donahue, Watson noted that Peter was "more interested in the structure of Nina, Perutz's Danish au pair girl, than in the structure of myoglobin." Crick, too, felt that the young Pauling was "slightly wild," but still the officemates hit it off immediately. According to Watson, Peter's presence meant that "whenever more science was pointless, the conversation could dwell on the comparative virtues of girls from England, the Continent, and California." Watson and the young Pauling even made a point of visiting The Rex art house cinema together to watch the 1933 romantic film *Ecstasy*, which Watson referred to affectionately as "Hedy Lemarr's romps in the nude."

Women aside, Peter was most concerned by the day-to-day troubles that were typical of English life in the early 1950s. He wrote to his mother about the absence of a bathtub in the small, cold, damp room that he now inhabited, and complained about the space's perpetual lack of sunlight. He did praise his fortune at having scoured London to find a suitable teapot, and he requested that Ava Helen kindly make him a pair of curtains for his window, which she happily obliged.

In letters to his father, Peter preferred to talk about cars, or his recent dinners with the Braggs and their daughter Margaret, rather than his own research pursuits. Linus, on the other hand, was immediately curious about the intellectual climate at the Cavendish and was especially interested in the work of Francis Crick, who a year earlier had been part of a collaborative effort to develop a theory of mathematical representation for X-ray diffraction that was fast becoming a standard in the field.

That same previous year, 1951, Pauling had bested Bragg and the physical chemists at Cambridge in becoming the first to publish the alpha helical structure of many proteins. But despite the desire prevailing at the Cavendish to eventually beat Pauling at his own game, Watson and Crick had been warned to keep away from the study of DNA by the head of the lab. Bragg knew that Maurice Wilkins and Rosalind Franklin, of King's College London, were already working on the problem using Franklin's photos and crystallographic calculations of the A and B forms (low and high hydration levels, respectively) of DNA.

Wilkins and Franklin's work was proceeding slowly however, and Peter Pauling and Jerry Donahue — another Caltech graduate now stationed overseas as a post-doc — were both in regular communication with Linus Pauling. Once casually conveyed, these contacts provided Watson and Crick with insight into what was going on in Pasadena. In his correspondence, Peter joked about the mounting competition between Caltech and the researchers at the Cavendish and King's College. "I was told a story today," he said to his father. "You know how children are threatened 'You had better be good or the bad ogre will come get you?' Well, for more than a year, Francis and others have been saying to the nucleic acid people at King's, 'You had better work hard or Pauling will get interested in nucleic acids.'"

While Watson and Crick urged Wilkins to provide them with Franklin's images and calculations so that they might model the structure themselves, Peter stoked the fires of their urgency, assuring them that his father was no doubt only moments away from solving the problem. Donahue was equally convinced: for him, Linus Pauling was the only scientist likely to produce the right structure.

In December, the fate that Jerry Donahue and Peter Pauling had been predicting seemed to come true: a letter from Linus to his son claimed that he had indeed determined the structure of DNA. The letter gave no details, simply confirming for Watson and Crick that Pauling and his Caltech partner Robert Corey had somehow solved the problem. Watson later recounted his colleague's distress in hearing this news, recalling that Crick "began pacing up and down the room thinking aloud, hoping that in a great intellectual fervor he could reconstruct what Linus might have done." But it seemed to be too late. Pauling's DNA paper was set to appear in the February 1953 issue of *Proceedings of the National Academy of Sciences*. In all likelihood, it would be time to move on to new projects.

*

With winter break coming fast and Pauling having apparently solved the structure of DNA, Watson and Crick extinguished any hope of modeling their own structure. Eager to take advantage of a few days off, their Cavendish officemate, Peter, headed for the continent in the company of a friend whom he described as "a mad Rhodes scholar" who had "wooed" him with his "insane plan" for exploring Europe.

On this trip, which was indeed ambitious, Peter visited Munich, Vienna, Linz, Brussels, Frankfurt, and Bavaria, hitchhiking his way from location to location. Crossing Germany, Peter saw neighborhoods still littered with the rubble of the Second World War, alongside industrious people struggling to rebuild. His mode of travel, he confessed to his mother, had seemed a better idea when its low cost was his only consideration. In person, however, spending several hours standing in or walking through the snow had a way of changing one's priorities.

Nonetheless, the whole escapade proved a romantic adventure for the young Peter Pauling. He spent Christmas Eve in a gas house belonging to the director of an iron company somewhere in Leoben, Austria. Resting there and watching the snow fall, he wrote again to his mother, "I look out the window to the lovely white mountains. It is grand. Considering the possibilities, Christmas and your birthday [Ava Helen was born on December 24, 1903] could hardly have been spent in a nicer place. Considering impossibilities, I can think of places where I would much prefer to be. Sometimes it is sad to grow up."

With the arrival of the New Year, the Cavendish researchers put their skis away, shook the snow from their coats, and resumed their work. It wasn't long into the term before Peter learned, from two letters received in February, that his father was, in fact, having difficulty with some of the van der Waals distances hypothesized to be near the center of his DNA model. In response — and almost as an afterthought — Peter casually asked his father for a manuscript of the DNA proposal, mentioning that his coworkers in Max Perutz's unit would like to give it a read. Upon receiving the paper, Peter promptly revealed to Watson and Crick that the Pauling-Corey model was a triple helix, a concept similar to one that Watson and Crick had developed themselves — and rejected — back in 1951.

This moment was a major turning point for Watson and Crick, who only then realized that they still had a chance to discover the structure before Linus Pauling. That said, what followed may not have been quite the race that it was made out to be after the fact. Peter, at least, did not see it that way, and the casual manner in which his father interacted with him (and with others at the Cavendish) seems also to belie such a dramatization.

Near the end of February 1953, while wishing his father a happy birthday, Peter noted that his office still felt that Linus's structure required sodium to be located somewhere near the oxygens, whose negative charges would have to cancel out to hold the molecule together. "We agree that everything is a little tight," he wrote, referring to the small atomic distances between Pauling's three polynucleotide chains with phosphate groups in the middle.

As communicated in an earlier letter to his son, Linus had already identified these structural arrangements as a weakness of the model, and he was in the midst of attempting to correct the issue. Peter confided to his father that, at that time, the Cambridge office had nothing better to offer. He added simply that "We were all excited about the nucleic acid structure," and concluded with many thanks for the paper.

In response, Pauling asked for updates on any progress that Watson and Crick were making with their own model, and casually requested that Peter also remind Watson that he should arrive at Caltech for a scheduled proteins conference by September 20th. Peter clarified only that the Cavendish group had successfully built the Pauling-Corey model and that Watson and Crick had then discarded it, becoming very involved in their own efforts and "losing objectivity." It would be up to them, Peter said, to communicate the details of their structure. Shortly thereafter, Watson and Crick sent a letter to Linus Pauling, outlining their structure and including a typescript of the short article that they had submitted for publication in *Nature*.

It has been well-established that Pauling and Robert Corey made basic errors in their own modeling of the structure of DNA. But in March 1953, having no knowledge of the X-ray crystallographic photographs of DNA that had been taken by Rosalind Franklin at King's College, Pauling felt bewildered by the certainty with which Watson and Crick had rejected his triple helical model. Upon learning its details, Pauling agreed that the double helix model was at least as likely, and he considered it to be a beautiful molecular structure, but he could not understand why his own structure was being ruled out entirely.

Crucially, Pauling did not believe that any X-ray evidence existed that proved that the phosphate groups might somehow be located on the outside, rather than in the core, of the DNA molecule. Pauling did not believe that this evidence existed because he hadn't seen it yet; Watson and Crick had. Indeed, from the point of their realization that Pauling had modeled the structure incorrectly, Watson and Crick worked fervently to once again convince Maurice Wilkins to provide them with Rosalind Franklin's data. In one instance during this period, while in Crick's basement dining room, Watson and Crick discussed the feasibility of redoubling their efforts to model DNA while Peter, casually eating biscuits and sipping tea at the table, offered that if they didn't do it soon, his father would take another shot at it. After the embarrassment of a failed attempt, he assured them, Linus Pauling was a strong bet to get it right the second time around.

Within a month's time, and with Rosalind Franklin having left his lab, Wilkins finally consented to providing Watson and Crick with all of the relevant data that they had requested. This proved to be the final piece that the duo needed to build their correct structural model of DNA. While all of this went on, Linus himself was seemingly unconcerned by any "race" for the structure of DNA. In fact, the only racing on his mind was a jaunt across Western Europe in a new sports car.

With Watson and Crick frantically working to unravel the secrets of DNA before Linus Pauling beat them to it, Pauling himself was debating the virtues of British, German, and Italian motor vehicles. Preparing for multiple trips overseas and in the market for some new wheels, Pauling's plan was to select a car while in Europe for the spring Solvay Conference, and then to actually pick it up in August, when he and Ava Helen would return for the International Congress of Pure and Applied Chemistry, held in Stockholm and Uppsala. The couple would then tour the continent in style before returning to the United States on a Scandinavian freighter and driving across the country from either New York or New Orleans.

While Peter advised his father that a Jaguar Mark VII was absolutely the best buy of the season, Linus expressed a preference for the slightly more modest Sunbeam-Talbot convertible. Peter countered with an Austin A-40 Sports 4-Seater, and Linus finally agreed to have Peter look into purchasing the car on his behalf and scheduling a delivery of sorts. Seeing that his father was finally taking the bait, Peter attempted to spring a trap: "Might you be in need of a chauffeur, mechanic, linguist, travelling companion, navigator, break repairer, tire changer, witty conversationalist etc. on your trip next summer? I know just the fellow. Good friend of mine."

As the end of March rolled around and the Solvay Conference approached, Pauling alerted his son to the fact that he had not yet made hotel reservations or, really, any plans for his visit to Cambridge. This responsibility he delegated wholly to Peter, who was somewhat distracted at the time, writing of the blue sky and sun that had finally begun to break up the English winter gloom, and announcing with pride that he had gone to two balls in one week, getting along quite well with the Scandinavian women. "As a sensible young American, I stand out in this town of pansy Englishmen," he declared with impunity.

But when Linus finally arrived at Cambridge in April, he found his son's sensibilities to be somewhat lacking. Peter had in fact not made the requested hotel reservations, and while campus accommodations were fine for the son, they were not so wonderful for the elder Pauling. As Watson later joked, "the presence of foreign girls at breakfast did not compensate for the lack of hot water in his room."

When the moment of truth finally came, Peter and his father strode into the Cavendish offices to see the model that Watson and Crick had constructed. Upon inspection, Linus repeated the assessment that he had given to his son earlier: the structure was certainly possible, but to be certain, Pauling would need to see the quantitative measurements that Wilkins had provided. By way of response, Watson and Crick produced "Photo 51," Franklin's now-famous image that enabled crucial measurements of the B-form of DNA. Presented with this evidence, Pauling quickly conceded that Watson and Crick had solved the problem. Later that night, the Paulings, together with Watson, had dinner with the Cricks at their home at Portugal Place to celebrate. To quote Watson, each "drank their share of burgundy."

So was it a race? And if so, what was Peter Pauling's role? Was he an informant? A double agent? Or merely an unwitting accomplice, ignorant of the full implications of his actions?

In trying to answer these questions, it is important to emphasize that, for Peter, the "race for DNA" had never been a race at all. His father, he believed, was only interested in the nucleic acids as an interesting chemical compound. Linus Pauling clearly didn't attack the structure with the same tenacity as Watson, in particular, who regarded the genetic material as the holy grail of biology, the secret of life. As Peter would write two decades later in *New Scientist*, "The only person who could conceivably have been racing was Jim Watson. Maurice Wilkins has never raced anyone anywhere. Francis Crick likes to pitch his brains against difficult problems... For Jim, however...the gene was the only thing in life worth bothering about and the structure of DNA was the only real problem worth solving."

In 1966, Watson, then in the process of writing his book on the discovery of DNA, *The Double Helix*, sent Peter an early draft. His concern, he explained, was that he accurately portray Peter's role in the entire affair; that, and he didn't want Peter to sue him for defamation. Peter laughed and told his old officemate that he thought it was a very good book; certainly very exciting. He pointed out, though, that Watson should ask Linus Pauling for an agreement not to sue him, too. After all, Peter said, "He has more experience than I do."

*

In the first months of 1953, with the DNA scramble surrounding him, Peter was mostly concerned with the English weather. He had been at Cambridge since the previous fall, and in that time had seen a mere two full days of sun. He was now officially fed up. His father, by contrast, was mostly concerned with finishing his most recent edition of *The Nature of the Chemical Bond*, for which he had often solicited Peter in the past as a source of example problems and solutions. But as he was now beginning his graduate research, Peter was too busy to provide much assistance this time around.

Instead he was mostly occupying himself with a camera developed by Hugh E. Huxley, a molecular biologist studying the physiology of muscle with Max Perutz's Medical Research Council (MRC) Unit of Molecular Biology at Cambridge. Taking pictures of fibrous and globular proteins — beginning with insulin and tropomyosin — Peter applied the Cochran-Crick theory, with the goal of determining the helical structure of these protein molecules. This inquiry was, in principle, made possible by Linus Pauling's work from less than a decade prior.

Since 1947, when the MRC unit was founded by Sir Lawrence Bragg, Perutz and John Kendrew had endeavored to use X-ray crystallography to determine the molecular structure of hemoglobin in sheep. By the time that Peter arrived at Cambridge however, hemoglobin had proven to be an untenable object of study, and Kendrew's focus had shifted to myoglobin. Whereas hemoglobin is found mostly in the blood, myoglobin is generally found only in muscle tissue. Both are proteins that carry oxygen to cells. Problematically, myoglobin is one-fourth the size of hemoglobin, meaning that it was too small for the era's X-ray analysis techniques.

To solve this issue, sperm whale myoglobin was used in hopes that the molecular details of the larger, oxygen-rich proteins of a diving mammal would be more observable with the tools then available. "Stranded whales are the property of the Queen," Peter explained to his father in a discussion of this work, "but we have an agreement with her to get a piece of meat if one comes ashore." Alas, though availed of samples from beached whales in the United Kingdom and countries as far afield as Peru, Kendrew could not render the X-ray diffraction patterns with complete certainty.

In 1953 Perutz realized that by comparing the diffraction patterns of natural whale myoglobin crystals to crystals soaked in heavy metal solutions — a procedure called multiple isomorphous replacement — the positions of the atoms in myoglobin could be more accurately determined. Accordingly, Peter was tasked with making countless measurements in support of this effort.

Peter wrote to his father often over the next two years as he struggled to complete the project, which was the focus of his PhD. In particular, Peter asked for advice on how one might best get heavy metal atoms onto myoglobin, detailing his attempts to use everything from saltwater to telluric acid, which was used to produce salts rich in metallic contents, such as the element Tellurium.

And while there were many reasons why Peter's work proceeded slowly, among them were his knack for keeping things entertaining. Shortly after Watson and Crick's discovery of DNA for example, he fabricated a letter of invitation from his father to Crick, requesting Crick's presence at an upcoming conference on proteins at Caltech. "Professor Corey and I want you to speak as much as possible during the meeting," the impostor Pauling said to Crick in the fake letter, even urging him to consider lecturing at Caltech as a visiting professor. Linus Pauling had appeared to sign the letter himself, his signature skillfully forged. The letter proved so convincing that Crick actually replied, accepting the invitation to speak at the conference.

But before long, it became apparent that the entire communication was, in fact, a practical joke, mostly because Lawrence Bragg, the director of the laboratory where Crick himself worked, was scheduled to speak at the conference in the same time slot that the fake letter had proposed for Crick. Were it not for this glitch the deception might have gone even farther, since upon seeing his son's forgery, Linus himself was almost convinced that he had written the letter and simply forgotten about it. Ever a stickler for the details however, Pauling noticed a grammatical error in the document that he would never have made, and deduced the letter as having been authored by his mischievous son. For this transgression, Linus subtracted a five-pound fine from the $125.00 check that he sent to Peter each month.

*

The year 1954 proved tumultuous for Peter. For one, Jim Watson had left for Caltech, and Peter lamented that his absence was felt, as he was "a positive force, albeit a bit conceited" when it came to social dynamics in the lab. At the same time, Peter's sister Linda was preparing to move to Cambridge, where her father hoped that Peter might help her find lab work assisting with crystal structure determinations. (Linda was quite interested in mathematics.) His sister's imminent arrival excited in Peter visions of European exploration, and especially of skiing.

But while Peter dreamed, serious matters were afoot at the Cavendish Laboratory, whose director, Lawrence Bragg, was planning to resign to take a position as head of the Royal Institution in London. Meanwhile, the lab's incoming director of physics, Nevill Mott, was widely known to be of the opinion that the unit's increasing focus on biology needed to be redirected. Amidst these changes, John Kendrew was worried that the MRC unit that he and Max Perutz headed might be kicked out of the lab, or even the department entirely.

The uncertainty both distracted Kendrew from Peter's lack of progress on his myoglobin work and, in retrospect, made Peter's lack of enthusiasm for the topic all the more glaring. Indeed, while Kendrew was worried about the future of their research, Peter was writing to his father that he was unconcerned about Mott's approval. Rather, as was so often the case, Peter's main preoccupation was his vehicle, this time a 1930 Mercedes Benz open touring car — described as "18 feet long and mostly engine" — that Peter was now cruising in for special occasions like the May Ball at Peterhouse. Peter's older brother, Linus Jr., had forwarded him money to purchase the car, hoping that it would be affordable to rebuild the engine. When the cost of doing so turned out to double his investment in the vehicle, Linus Jr. decided it more expedient to simply let his younger brother have the car.

Linus Jr. and Peter formed a strong bond during Peter's years at Cambridge, a time period where Linus Jr. and his wife Anita made a habit of travelling around Europe during the summers. This closeness marked something of a renewal of the brothers' relationship since they had seen little of each other during their more formative years and, as children, had few interests in common. Now, cars in particular emerged as an area where the two could share their exuberance. Linus Jr. reflected later that, on those trips abroad, he and his wife enjoyed Peter tagging along — his vitality, beaming smile, and friendly nature made him the life of any party.

But this was clearly only one side of Peter Pauling. Privately, he admitted to his mother that he often felt unsure of his path in life, and was not certain of his ability to meet the challenges of his PhD program. He often wondered whether or not he would be better off simply teaching chemistry, or helping to write his father's textbooks. These bouts with gloom were contrasted by sudden and excited turns to sociability. Linus Jr. would later point out that their paternal grandmother — Linus Pauling's mother, Belle — was possibly manic depressive, and was reported to have died in a mental hospital. This, he believed, was likely where Peter had inherited his own emotional instability, and it was during his stint in Cambridge that manic depressive symptoms started to manifest most clearly.

Linus Pauling's frustration with Peter's hoax "Francis Crick letter" had faded by the time the entire family met in Stockholm for the 1954 Nobel Prize ceremonies. It was there that Pauling was to receive his highest honor to date, the Nobel Prize for Chemistry, commemorating his work on the nature of the chemical bond. After a frustrating battle to receive government permission to leave the country — Linus's political activities were causing him problems with the Passport Office — the

Pauling family flew to Copenhagen where they met Peter and Linda. By then, Linda had taken up residence in the basement room that her brother had just left at the "Golden Helix," as the Crick home on Portugal Place was now known. Once arrived, she worked for a time as Francis and Odile Crick's au pair.

Watson returned to the Cavendish in 1955 to find the MRC unit on the verge of being squeezed out by Nevill Mott. Finally registering this threat, Peter began to panic, writing to implore his father that he vocalize his positive impressions of the unit's work and that he recommend that the group be allowed to continue their research there. At the same time, Peter applied for a post-doctoral fellowship grant from the National Science Foundation, hoping to solidify the standing of both himself and the group by bringing additional research money into the lab.

As it turned out, Peter's maneuver worked: he received the grant, and this was no doubt a boon to his position at a crucial time. It did little to help him in his research, however. He continued to struggle with myoglobin and, increasingly, he placed his fading hopes squarely upon the idea that mercuric tetraiodide ion crystals might be a better candidate for the sorts of analysis that Kendrew and Perutz were beginning to doubt he could complete.

As the final year of Peter's program dawned in fall 1955, the frequency of his drives about the grounds to impress the ladies dropped to what Jim Watson considered a startlingly low level. Perhaps realizing the "do or die" position that he was in, Peter seemed to be redoubling his focus on finishing his degree. During this same period, Peter had begun seeing a young woman by the name of Julia, who was a student at a nearby all-women's school. Watson, curious about the situation, queried several women that he knew from the school, but most were silent, and Julia herself became conspicuously absent as the New Year drew closer.

Meanwhile, Peter's father had been working to prepare his son for life after Cambridge, offering him an appointment in the Caltech Division of Chemistry and Chemical Engineering as a Research Fellow focusing on the crystalline structure of globular proteins. As Pauling wrote, "We have a real need here for someone who has had the sort of experience in taking X-ray photographs of crystals that you have obtained. I think our effort to determine the complete structure of a crystalline globular protein is going to be successful, and that you might like to be associated with the successful effort."

Peter did not respond immediately, taking about a week to think about the proposal. It may well be that he was simply overwhelmed by both the work to be done and the festivities to be had during his final months at Cambridge. Plus, it seemed that the job his father had offered likely would be waiting for him as soon as he had finished his program in the U.K.

Few had seen much of Peter in the run-up to a dinner party that Jim Watson and Linda Pauling were planning as a practical joke. In response to a rumor that Watson and Linda were seeing one another, the two decided in good fun to host a get-together, thus fanning speculation into a frenzy by implying an impending announcement that, in fact, was never to come. Peter was invited and did show up, but much to the surprise of the hosts, he was not his usual grinning, charming self. Instead, he seemed sentimental and full of a solemn interest in his friends' futures at Cambridge. Watson and Linda later realized that, on this particular evening, Peter was wrestling with an especially weighty issue: he was soon to become a father.

*

A letter sent by Peter's parents in early 1956 concluded with an expression of excitement: Linus and Ava Helen would be visiting soon and would look on with pride as they witnessed their son receiving his Cambridge PhD. In his response, Peter explained that this day, sadly, would never come. Though he felt that she was a "clever, delicate, and lovely girl," Peter had not made Julia an "honest woman," and for this he would be sent down from Cambridge and not allowed to take a degree. By extension, this also meant that he would not qualify for the position that his father had offered him at Caltech.

When he learned of this situation, John Kendrew suggested that Peter might be able to transfer both the remainder of his fellowship with the National Science Foundation, and also the completion of his doctoral research, to the Royal Institution in London, where Sir Lawrence Bragg — his old program director at the Cavendish — was now director of the Davy-Faraday research lab. By then, however, Peter had decided to marry the mother of his child, and arrangements were quickly made by Linda Pauling for a quiet civil wedding that was out of the spotlight and not attended by either of the Pauling parents.

Peter and Julia were married on March 13, 1956 at the Cambridge Register Office on Castle Hill. Peter's bride was given away by her father, and with no family members other than Linda present, Peter's sister acted as the sole adjudicator of the Pauling family's approval of the union. Kendrew, Peter's Cambridge advisor, stood with him as his best man. Following the wedding, a reception was held at Kendrew's home at Tennis Court Road, after which Peter put on his trademark grin and, with Julia, vanished in a new Porsche. Before the year was out, Linda Pauling, struggling financially and burdened by an expired work visa, returned to Pasadena.

Between 1957 and 1959, Kendrew and Perutz successfully modeled the molecular structure of myoglobin that Peter had been working on. In this, the Cavendish once more beat Caltech to the punch, as the position that Linus had offered to Peter was meant to contribute to a similar problem. Myoglobin was the first ever protein to have its atomic structure determined, and Kendrew and Perutz shared the Nobel Prize in chemistry for this achievement in 1962.

*

Journeying on their honeymoon through the caves of northern Spain, Peter and Julia arrived at a small fishing village and made camp. His beard full and with hair grown to nearly his shoulders, Peter sat on the beach, scouring pots. Meanwhile, Julia watched the water, contented by the meal that she had just prepared for her new husband. She had always loved the sea, saying as much in her letters to Linus and Ava Helen, her new parents-in-law.

Julia had been a bright student at Cambridge. An avid reader of French and German literature, she was once hailed as the year's best student at Girton College and had received the highest marks possible in her first-year examinations, an achievement that surely would have impressed the new in-laws. Given the circumstances of their marriage however, Peter and Julia had their work cut out for them in attempting to smooth things over with both sides of the family. As she attempted to do so in her communications, Julia was especially complimentary of the Paulings' new property at Big Sur. In particular, she swooned over the "heavenly" view of the Pacific Ocean, as observed from the coastal bluffs of California.

Upon their return to London, the pair moved into their new home in Clapham, which Peter described as "an ugly Victorian suburban house that ought to be quite pleasant." Many things changed for Peter as he settled into domestic life. With the help of his mother and younger brother Crellin, he shipped his Mercedes back to the United States where it would eventually be sold. He had likewise traded in his most recent automotive triumph, the Porsche, upon his and Julia's return from their honeymoon. The proceeds from these sales and trades were used — with added financial help from Peter's parents and older brother Linus Jr. — to purchase a new home for the family at Lansdowne Road.

By July 1956, Peter and Julia were thinking of names for their baby, with Peter being "most uncooperative about this business," according to his wife. In her letters to Ava Helen, Julia noted that every time she suggested a reasonable name, Peter demurred, offering alternative suggestions like "Gregorio" and "Plug-up," the latter a character from what Julia considered a "pointless space-fiction strip cartoon."

When the baby finally came on August 22, the pair had settled their differences. Peter Andrew Thomas Pauling, to be called Thomas, was born that summer, to be followed by a younger sister, born February 5, 1960. Again there seems to have been some measure of disagreement about a name, with Peter first announcing his daughter to his parents as "Esmiralda Ermitrude."

That name didn't stick however, and within a month, Peter was writing to his mother and father that their new granddaughter, Sarah Suzanne, had begun to smile and sleep all night. It was one of many letters in which Peter expressed joy at being a father. Within five years, he would excitedly report that Sarah was reading bedtime stories to him, rather than the other way around. By this time, young Thomas was at the top of his class as well.

Peter and Julia enjoyed a great deal of support from friends and family during these early years. Typically, the couple would spend the Christmas holiday season with Julia's parents in the north of the U.K., while the Pauling family would usually visit at various points throughout the year. Occasionally, Peter's sister Linda and her husband Barclay would see the young family, bringing their twin boys "Barkie" and "Sasha" in tow. Linus and Ava Helen often came through London while on European trips for conferences, delivering comfort items from the States as well as more important cargo, including the polio vaccine.

Thomas and Dorothy Hodgkin stopped in regularly, as did the Cricks and the Bernals. Joy and J.D. Bernal gave Thomas his first toys and provided the cake for Sarah's first birthday party. The Hodgkins offered Peter and Julia their old baby bath, and Francis and Odile Crick passed along some hand-me-down clothes. "So far," Peter joked, "the entire cost of the baby has been one box of chocolates for the nurses."

Even Jim Watson dropped in, meeting young Thomas, who loved to turn all the knobs on a sprawling electronic gramophone that Peter had pieced together from spare parts. The room was a hopeless mess, Watson noted, and surely the bane of Julia's existence. On top of that, Thomas's interventions generally scrambled the music until it was incomprehensible.

Buoyed by a little help from his friends, Peter's career took a positive turn as well. Lawrence Bragg had agreed to take Peter on at the Royal Institution for three months, in order to allow him to finish his degree. When those three months turned out to be not enough time, Bragg and Kendrew conspired with their colleague Jack Dunitz to obtain for Peter a position as a research student, working under R.S. Nyholm at University College London. Once there, Peter would be allowed to continue his education while simultaneously collaborating with Dunitz to complete his research.

The arrangement worked. Peter switched the focus of his dissertation from protein crystallography to inorganic molecules, using X-ray diffraction to verify configurations of a halide compound, $NiCl_4$. Peter likewise worked with a number of other transition metals, performing stereochemical experiments to determine their atomic structures. At the same time, he also began working with his father to develop a theory of the molecular structure of water, a subject on which he had spoken at a meeting of the Royal Society in 1957. After the two Paulings developed their theory, which postulated a random dodecahedral structure for liquid water, Peter became quite prolific. Throughout the late 1950s and 1960s, he published just over 30 papers, including 14 in 1966 alone.

He also became much more active in the field, flying often to the United States for meetings of the Crystallographic Association, as well as other conferences in locations from San Diego to Denver to Pittsburgh. Having completed his PhD in 1959, Peter was immediately offered a lectureship in Chemistry at University College London. And though he continued to muse in his letters to Ava Helen that he really didn't want to do chemistry forever, he quickly accepted the position.

At long last, Peter seemed finally to be stepping out from his father's shadow. Key to this was his own realization that he was not, and could never be, the chemist that Linus Pauling was. Instead, Peter began to focus his efforts on computers and other electronic systems valuable to the lines of chemical research that he had been pursuing. Among the rash of papers that he published in 1966 was "A Program for the Use of Large Computers for Crystallographic Problems," which appeared in the *British Journal for Applied Physics*. Here, Peter was finally in his element, working at the forefront of a field that was swiftly changing, engineering devices by hand, and building complicated electronics systems such as a "one dimensional diffractometer" for X-ray crystallography — or what Peter called an "automatic gadget" — from plug-in logical blocks.

Peter took important first steps forward in this new line of research when he ordered a computer and electronic parts that he thought would be necessary to produce a copy of the state-of-the-art diffractometer and visualization systems then in use at Oak Ridge National Laboratory, the American research center founded in 1942 as part of the Manhattan Project. Funded by a Public Health Service grant, his system-in-progress deployed an ex-military scope equipped with a pre-amplifier, a Schmitt trigger, a monostable pulse generator (used to trigger the scope), and a Sherwood FM tuner that he had acquired from Linus Jr. The tuner in hand, Peter spent almost a year tracking down its circuit diagrams so that he could most effectively cannibalize it to support his cobbled-together atomic measurement machine.

Once completed, not only did Peter's device work, it worked marvelously. By May 1968, his computer and the program that ran it were making thousands of minute measurements per week. Indeed, the apparatus was used to determine the structure of five compounds in a ten-week period; a volume of calculations, as Peter pointed out, that was visually represented by four miles of punched paper tape that the computer had to read in producing the work. This huge success stood in stark contrast to Peter's years at Cambridge, where he had struggled mightily to determine the structure of a single compound.

With his machine, Peter was attempting to make University College London technologically competitive with an institution that had received major support from the U.S. Atomic Energy Commission during the height of the Cold War. Astoundingly, he accomplished this goal using, to a large degree, spare parts. Later, Peter would use the measurements from his device to improve the Caltech method for drafting pseudo-perspective drawings of molecular structures, producing instead Third Angle Projection-style drawings of atoms and their bonds.

As his successes mounted, he was promised a lab that would be four times larger, and was elected President of the Chemical and Physical Society of University College London.

Behind the scenes however, Peter was struggling to balance his career with his family life, and was plagued by personal demons. Ever since leaving Cambridge a decade earlier, his mother had been worried about his mental health, urging him to see a psychiatrist about his manic depression. Over time, this view came to be shared by a growing number of friends and family. But burdened as he was by the competing forces of a new wife and children, the completion of his degree, and the press of research and professional obligations, there never seemed to be a good time.

At one point, Linus Pauling became so concerned for the welfare of his grandson, Thomas, that he offered to arrange for the boy to live in Pasadena for as long as might be necessary for Peter's domestic situation to stabilize. Peter responded that he was far too busy writing his thesis and preparing lecture courses at University College London to fly Thomas to New York. A few months later, he revealed that Julia was pregnant with their second child.

As time passed, the growing strain on Peter and Julia's marriage became palpable to those who knew and loved them both, and by 1961 Peter had suffered a serious breakdown, confiding to his parents that he was finally and earnestly trying to see a psychiatrist, as his bouts with sadness had become "uncontrollable." Peter's lament seemed, at times, to mirror the dark geopolitical climate of the 1960s. After John F. Kennedy's assassination, Peter wrote to his mother that he was "stricken" by the President's death. The optimism of the Kennedy years had led him to think that "ordinary mortals" might "rest a little easier" under the vibrant president's leadership. "Now," he admitted, "I fear it is back to the struggle."

And as the decade moved forward, new struggles emerged. By 1967, Peter and Julia had agreed to a divorce. Peter subsequently moved into a flat in a dodgy area of London — St. John's — where he shared his new space with a painter. The flat was later robbed, leaving Peter without most of his clothes and jewelry, as well as his radio, a loss that he lamented. ("I used it all the time," he wrote, "to fill up the empty holes in my head when I am alone.") Likewise stolen was a pot that his sister Linda had given him for Christmas. He wrote to her that he missed this item the most, as it meant more to him than anything else that was taken.

Linus Jr. came to London to visit his brother during this time, and ultimately left the scene with feelings of both worry and relief. The worry came from the fact that Peter, by his own admission, was drinking and smoking far too much. On the other hand, Linus Jr. felt a measure of relief that his brother had finally done what he thought was right for his children: leaving the family home at Lansdowne Road to Julia, Thomas, and Sarah.

*

It was the 1970s, and Peter Pauling was studying the molecular arrangement and physiological effects of hallucinogens. The field of psychopharmacology, crucial to psychiatric treatments of mental disorders, was at the time almost brand new. Only in the 1960s did most physicians begin to consider the potential of psychoactive pharmacological treatments in treating mood disorders and neurologically based physical ailments. The role that different substances played in the alteration of brain chemistry, their influence on synaptic changes, and the modifications in nervous response that they could bring about were all still poorly understood.

Peter's work modelling the structures of different neurotransmitters and psychotropic drugs was part of a larger, late 20th-century effort to address this gap in scientific knowledge. Many doctors of the era were studying drugs like mescaline, psilocybin, and lysergic acid (LSD). Often, these researchers self-administered as part of the experimental endeavor.

Peter was no exception. His first experiences with LSD came in 1962, when he began receiving small doses regularly as a treatment for his manic depressive symptoms. And while he generally found the results to be agreeable, his brother Linus Jr. was suspicious. A psychiatrist and graduate of Harvard Medical School, Linus Jr. advised that the treatment was not in favor in the United States, arguing that it was often prescribed by "unprincipled" doctors. Linus Jr. would later reflect that lithium as a treatment should have been offered to Peter at a much earlier point, and that it might have more seriously and effectively addressed his condition.

This compelling new research topic in hand, Peter dove into biophysics and psychopharmacology, attending conferences in Moscow and Vienna that inspired him to study drugs and molecules active in the cholinergic nervous system. His electronics work from the 1960s also continued to expand in tandem with his new scholarly focus. The grant that he had received to build his own computer and diffractometer served as the kernel for a grander plan to develop a data collection facility to research the nature of acetylcholine and other substances.

Indeed, Peter's stated aim was nothing less than to become "the world's expert on the structure of drugs," a goal he pursued with vigor throughout the 1970s. From 1966 — when the Medical Research Council began a systematic investigation and correlation of nervous system pharmacology — through 1979, Peter published no fewer than 48 papers in the field. His major works came to print in 1970 with "The Conformation of Molecules Affecting Cholinergic Nervous Systems"; in 1972 with "The Molecular Structure of LSD"; and in 1973 with two articles, "Neuromuscular Blocking Agents" and "The Conformations of Neurotransmitter Substances."

During this fertile period, Peter was supported by grants averaging about $100,000 a year, including funding to purchase a computer graphics display and to employ a programmer to assist in the development of an interactive graphics system for the study of molecular structures. He described the function of the new technology to his father with awe: "One just pulls up a picture of the molecule and wiggles it around until one gets a pretty view, and punches a button which sets up a file which consists of the entire job input file for another computer to draw the picture. One then rings up the other computer on the telephone and sends the file down to it!"

To date, no serious consideration appears to have been given to the impact that Peter may have had on the young field of psychopharmacology during his short but prolific research career at University College London. He himself saw his work as influential, claiming in 1993 that his goal had been to discover details overlooked by other researchers and lamenting that, in his view, scientists in his own field had, in fact, overlooked these discoveries for years.

Peter's personal life seemed to be improving at this point as well. Having moved out of the slum and into a new flat on Hornton Street, everything looked to be coming together. Early in the decade, Peter began dating a book publishing agent by the name of Bud. The two had originally met at a party in Cambridge years earlier, and before long, Bud and Peter were married, with Peter's daughter Sarah giving her father away at the ceremony. The couple subsequently moved in together and spent Christmas week in 1971 with some friends in a cottage in North Wales. Peter was enchanted by the place, noting that its Roman tracks and bridges inspired him to imagine someday owning a cottage of his own nearby.

But in stark contrast to pastoral dreams of retiring to the Welsh countryside, Peter's life had become a blur of work and travel. He flew to Helsinki, Copenhagen, Paris, and Sweden for conference after conference between 1970 and 1978. He also spoke, in 1974, at Peter Waser's Cholinergic Nervous System meeting in Zurich. (Waser had first outlined the role of cholinergic receptors in 1960, and is an early and influential figure in psychopharmacology.) Peter even received an honorary MD from the Karolinska Institute at Stockholm in 1972.

By now, Peter was also priding himself on his ability to keep what he called the "demon rum" securely in the bottle, though he likewise admitted that his was an ongoing struggle. By 1978 he had been hospitalized about five times — sometimes having himself committed — to receive psychiatric treatment for his depression as well as aversion therapy for his alcohol dependence. He took nicotinic acid and Antabuse to treat his symptoms, and adhered to a steady regimen of small doses of lysergic acid as well. When he was feeling well, Peter rewarded himself for all of his hard work with a new classic car; this time a 1938 Rolls Royce Phantom III with a 7-liter, V-12 engine.

Meanwhile, by 1972 Peter's ex-wife Julia had finished both a teacher's training degree as well as a bachelor's degree in Education. The children also seemed to be flourishing: Thomas had joined his school's rowing team, finishing in the top eight and winning a pewter cup in one race, and Sarah was beginning secondary school. The next year, Peter spent Christmas with Sarah and Thomas for the first time since 1966, after which his visits to both of them became more frequent, especially with Thomas. In 1979 Thomas joined his father for a trip to California to see his grandparents. Sarah accompanied Peter and Bud on a separate holiday that Peter admitted was an "eye opener" in showing him that he and his daughter could connect on a level that he hadn't been sure was possible.

Before long, Sarah was heading off to the University of Bristol, and Thomas was coming up on his final exams at Sussex. Much like his father, Thomas possessed a mind that was mechanically focused, delighting in daydreams of designing bicycles or someday becoming an electrical engineer. In 1980 Thomas took a position at an engineering firm working in "forecasting expenditure," while living in the same Hornton Street building as his dad. Sarah would later complete a bachelor's degree in Biology and a master's in Forensic Science, ultimately forging a career in pathology at the U.K. Home Office.

By the early 1980s, Peter had entered into a romantic relationship with another woman, Alicia, a librarian and classics instructor who lived in the same apartment building as he. Bud had moved out, and though she wished to maintain contact after their separation, Peter was less open to the idea. He preferred to continue to see Alicia, their "on again, off again" relationship percolating for many years before the two finally married. And though they took their time before sealing their vows,

Alicia pretty quickly became a crucial fixture in Peter's life, providing much needed support when tragedy repeatedly struck.

In 1981, Ava Helen Pauling, in failing health and suffering a recurrence of stomach cancer, decided against chemotherapy. She passed away on December 7th at the age of 77. Not long after, Peter lost his son Thomas, who, in 1983, died tragically at the young age of 26. The next year, Peter was admitted to the Queen Mary's Hospital burn ward in Roehampton, where he received surgery to aid in his recovery after unintentionally setting his mattress on fire in the middle of the night.

In 1986 Peter accepted an offer to take an early retirement from his job at the University of London. His academic career now concluded, Peter channeled his energies into fulfilling the vision that had first occurred to him in 1971: a small home in Wales. After many trips spent looking at properties, Peter and Alicia eventually came across a house that was adjacent to an isolated and run-down mill out in the countryside of Dyfed.

For Peter the mill itself was the real attraction. It was called Abergwenlais, and it was 250 years old. Its specs included two big rooms, five bedrooms, two wood stoves, and a leaky roof in need of repair. Parts of the facility still retained the original equipment, though the mill wheel itself, which likely had been 20 to 25 feet in diameter, was missing. Linus Jr. later described the property as follows:

> "Through the ground ran a stream, part of which was diverted into a sluiceway to motivate the wheel. Inside the mill was still the mechanism for transferring energy from horizontal to vertical grindstones, which had big doors on each side so wagons could go in underneath and get loaded up and then on out the other side. Only a small portion had been made barely habitable, with running water and a bathroom, and what passed for a kitchen. The Welsh countryside was beautiful. The place suited Peter."

Peter and Alicia married in 1991, taking a trip around the world as their honeymoon, and paying visits to Linus and Linus Jr. along the way. The next year, Alicia retired from her university job and sold her London flat, moving full-time into the house by the mill. But she continued to travel back and forth and remained academically active in her spare time. She also brought to the mill a number of cats that Peter considered "bloody awful things" for preying on the local birds.

It is clear, however, that Alicia's presence was supremely positive for Peter, both physically and psychologically. In a 1992 letter to his father, Peter admitted that, after years of struggle, he had finally found someone who would look after him. "I try not to complain, but just let things slide by," he said of pastoral life with Alicia, confiding that "it works quite well. She says she will be here more or less permanently by the end of September, but this may be a bit optimistic. It always has been."

Yes, it was optimistic; but only just. Alicia returned permanently to the mill a month later than planned, and the pair lived out their days there together among the old Roman roads, the green hills, the grey skies, and the deep winter snows.

*

Peter Pauling passed away on April 22, 2003. He was 72 years old. Before he died, he saw his daughter Sarah marry, and also witnessed the births of two grandsons, Isaac and Malachi. Over time, he likewise learned to recognize the ebb and flow of his manic and depressive phases, at points struggling to overcome insomnia and drinking too much whiskey or beer, and at others walking the country paths around the mill so giddy with delight that he felt he could not contain his joy.

In 1992 Jim Watson came to Wales to call on Peter and Alicia. Peter had recently seen a BBC drama depicting the discovery of DNA which, as he explained to his old friend, was not entirely accurate. When Watson asked what they had got wrong, Peter answered firmly that he had appeared only at the very end of the program, and that he showed up on screen driving a white Cadillac convertible. For a car man like Peter, being portrayed in such a vehicle was, apparently, an insult to his sense of personal pride.

Though thousands of miles apart, Peter remained in regular contact with his father. In 1992 Linus called on his son to ask his advice about what he should do with a collection of secret documents stemming from his years of involvement in the American war effort. As he looked through his files, Pauling had been unable to track down an apparently non-existent Navy patent for a substance, named "Linusite," that he helped to develop in secrecy in 1945. Similarly, he noted, his invention of the oxygen meter had presumably remained classified, as was a cone shell windmill that he designed in 1952.

Indeed, the elder Pauling had a personal safe full of records relating to such top-secret projects, and he had no idea which of them had been declassified. Now, at the age of 91, he wanted to unburden himself of these materials, one way or another. Wishing to help his dad out, Peter called a close friend of his from his undergraduate years at Caltech, Robert Madden, who was then working in the National Security Administration. Pauling's safe was subsequently inspected, and select material duly vanished into the hands of the federal government.

A year later, the conversation had turned toward the introspective. As his cancer spread and his health continued to diminish, Pauling lamented to his son that he had never much been there for him during Peter's childhood; had never thrown the baseball around the yard. His son responded in stark contrast, stating that he looked back on his childhood at Arden Road and Fairpoint Street with great fondness, adding,

> "Well, you did not play much baseball, but then neither did I. You did, however, lie on the side of my bed and taught me how to count in French. Later, when I was old enough to get out and about, you were often out to rescue me, either because I telephoned or Mamma was worried and sent you out to do a general search of the whole of Pasadena and surrounding environs."

In 1994 Alicia sent a letter to Linus on his birthday, saying that she and Peter were thinking of him, and about to toast him, as dinner time was drawing near. The drinking, she hastened to add, would be kept moderate, but the thinking had no limits. She concluded by writing that "Peter hopes to come over shortly — and so do I."

Less than six months later, Linus Pauling passed away. Peter's younger brother Crellin wrote to him after their father's death, heavy with grief. Peter, though, had been experiencing both the depths of depression and the heights of elation for decades. The lesson in all of this, he confided to Crellin, was that when one's mania had faded and the depression set in, one had only to hold on. Be patient and outlast it, for eventually change will come. In this, an entire lifetime of often difficult experience was summed up by Peter Pauling in three simple words: "Do not despair."

18 Patents

By Olga Rodriguez-Walmisley and Trevor Sandgathe

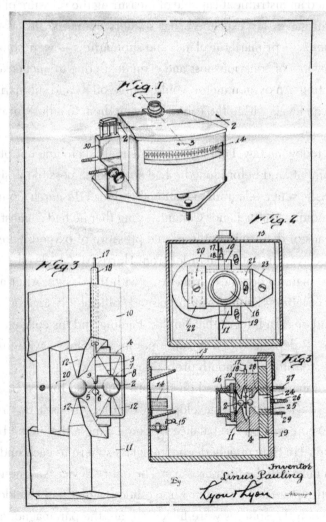

Diagrams of the Pauling Oxygen Meter, June 8, 1942.

Oxygen Meter

> "Have most promising method determination partial pressure oxygen.
> Best available post-doctorate assistant offered job elsewhere. May I
> hold him. Please telegram or telephone."
> — Telegram from Linus Pauling to James B. Conant, October 8, 1940

On October 3, 1940, Linus Pauling met with his colleague, W.K. Lewis,
in New York City. At this meeting, Lewis informed Pauling that the military needed an instrument capable of measuring the pressure of oxygen
in a mix of gases. He explained that soldiers operating in low-oxygen
environments — primarily airplanes and submarines — were sometimes
affected by loss of consciousness and even death due to unchecked oxygen depletion. An oxygen meter would enable pilots and submariners to
track oxygen levels within the cabin, allowing them to adjust for dangerous decreases.

The following day, Pauling began mentally sketching out plans for
the instrument, and before long he had struck on a possible design. On
October 8, he sent a telegram to National Defense Research Committee
(NDRC) administrator James Conant stating that he had a "most promising" means of determining the partial pressure of oxygen. Soon after,
Pauling received an unofficial order from Harris M. Chadwell, Conant's
right-hand man, to begin experimenting with the design. After another
exchange of letters, Pauling was appointed "official investigator" for the
project and given a budget that funded Pauling and his colleague Reuben Wood — deputized here as a temporary assistant — with materials
and equipment for a six-month probationary period.

The apparatus was based on the principle of a torsion balance, a
measuring device originally developed by Charles-Augustin de Coulomb
in 1777. Wood created the balance by connecting a tiny metal bar to a
quartz fiber. He then attached a hollow glass sphere to each end of the
bar, and a mirror to the fiber crossbar. The entire device was then strung
between the points of a standard horseshoe magnet and shielded by a
bell jar. When the spheres were filled with air, the paramagnetic forces

present in oxygen atoms would cause the dumbbell setup to rotate, twisting the quartz fiber. As it twisted, the mirror on the fiber would alter the angle of a reflected light beam, striking a photocell. The photocell readings would then register on a dial, in the process giving an approximate measure of present oxygen levels.

By November 1, less than a month after receiving the assignment, Pauling and Wood had constructed and tested a model. Though fragile and prone to decalibration, it worked. After presenting the prototype to government officials, Pauling was told that the meter would need to be usable in a wide array of environments including frequent acceleration and deceleration, tilting on all axes, and constant shock and vibration. Thus charged, the duo designed an adjustable support for the apparatus that allowed it to remain stable despite movement and shock. Shielding and damping techniques were developed too, allowing the meter to give accurate readings under moderate strain from outside forces.

There were, of course, setbacks. The quartz fibers were nearly invisible and required special tools to create and place. The glass bubbles used on the instrument had to be hand-blown but were so delicate that it took many tries — sometimes hundreds — to create a single perfect bulb. Supplies, too, were difficult to acquire. Liquids for damping, metals, and magnets all proved hard to find, further slowing the research process.

Though development was cumbersome and sometimes frustrating, it was clear by the summer of 1941 that Pauling's oxygen meter was a success. The NDRC, pleased with the work, renewed his contract, requesting that five additional meters be manufactured and distributed according to committee orders. Despite the production demands placed on his team, Pauling insisted that the design be further improved. In response, Wood suggested a prototype using two magnets rather than one — a new approach that allowed for a sturdier, more accurate model.

Amidst the creation of the Office of Scientific Research and Development (OSRD), an entirely new war research agency, as well as major changes in hierarchy and administrative procedure, Pauling's group worked for seven months manufacturing oxygen meters with little inter-ference from officials. But by early 1942, it was clear that the team could not keep up with growing demand for the device.

It was at this point that Pauling turned to Caltech staff member Arnold O. Beckman, a skilled instrument-maker who quickly accepted a contract with Caltech for the manufacture and distribution of the oxygen meter. To simplify the production process, Beckman built the world's smallest glass-blowing device to create the meter's bulbs. Through this and a few other innovations, Beckman increased his manufacturing capacity to nearly 100 units monthly — at least ten times what Pauling's team could have hoped to achieve.

For the remainder of the war, Pauling continued to oversee the production and distribution of the oxygen meter, and Beckman, with his refined manufacturing process, succeeded in equipping Allied forces with hundreds of the devices. Customized models were also provided to laboratories and government institutions in both the U.S. and abroad, and were instrumental in the development of life-support systems for both pilots and sailors.

*

In a 1940 letter addressed to the NDRC, Pauling stated that, in view of the circumstances, and because his desire was to be of service to the country, he was willing to grant the government a non-exclusive, roy-alty-free license covering the entire invention throughout "the period of national emergency," referring to World War II. He also expressed his desire that the NDRC decide who would be given the rights to the apparatus at the end of the war. But parallel to these wishes, Pauling was also interested in filing an application for a patent on his invention "inasmuch as it seems it will be of use in various fields other than that of national defense."

Irvin Stewart, secretary of the NDRC, wrote back and essentially told Pauling that, because he had created the invention after sign-ing a contract with the Committee, the government was entitled to a royalty-free license on the device not only during the war, but through-out the life of the patent.

In a letter to James Conant, written February 15, 1941, Pauling reframed his desire to focus on patenting the fundamental idea, "now that my oxygen meter will soon be put in use in other laboratories," rather than the actual device itself. In this, he mentioned the contract agreed to by the NDRC and Caltech, which stated that the Committee would have the sole power to determine whether or not a patent application should be filed. He also noted that "there are many uses to which the instrument might be adapted other than the original one."

Pauling received an answer from Irvin Stewart on March 28, 1941, in which Stewart advised Pauling to apply for a patent on all of his developments that antedated the contract between the Committee and Caltech. Pauling replied that it was only after attending a meeting of the National Defense Committee in Washington, D.C. on October 3, 1940, that he initially learned of the need for an oxygen meter, and it was from this meeting that his ideas stemmed. Pauling's desire to patent his idea was running into roadblocks, but the uniqueness of what he had created could not be denied.

Pauling's "Apparatus for determining partial pressure of oxygen in a mixture of gases" was unique for many reasons. For starters, it was both lightweight and tough. It also made use of the fact that oxygen is a strongly paramagnetic gas, which means that its magnetism does not become apparent until it is in the presence of an externally applied magnetic field. Only a few gases other than oxygen are paramagnetic, but they are less susceptible to magnetism than is oxygen. For this reason, the apparatus was valuable in determining the oxygen content of a mixture of gases, except where other paramagnetic gases — such as nitric oxide, nitrogen dioxide, and chlorine dioxide — were present.

Because Pauling's device was going to be used in war, the government wanted to limit the number of people who knew of its existence. The NDRC eventually granted permission for Pauling to reveal the nature of his invention to his patent attorney in Los Angeles, provided that he refrain from disclosing the nature of the invention to anyone else. When Mr. Richard Lyon, of Lyon and Lyon, Attorneys, requested information on the assembly of Pauling's invention in order to better research existing inventions like it, Pauling asked his collaborator Reuben Wood to provide an update. It is from this exchange that we learn a bit more about what made the device special.

Wood told Lyons that Pauling's device was novel in many ways. For one, Wood could not find any other reference to the use of the magnetic susceptibility of oxygen as a means of analyzing a mixture for it. The Pauling method was also unique in that the composition of the gas sample was not altered by analysis, and "the moving part of the device is actuated directly by the presence of the gas in the analyzing chamber."

A somewhat similar apparatus, designed by Glenn G. Havens, required a recovery period of three minutes after being jarred or after a gas sample reading; Pauling's only needed one second. Another major difference between the two devices was that Pauling's was portable while Havens' was immobile and fragile. Furthermore, Pauling's model utilized a permanent magnet instead of an electromagnet, which meant that his device weighed less. Also, the only source of electricity needed by the instrument was that required to operate a light bulb, which could be powered using a flashlight cell. All in all, Pauling's model was more efficient, more portable and more dynamic than any competing instrument. Wood believed that all of these attributes were patentable.

Pauling filed a patent application on August 23, 1941, after which he was promptly informed by the Department of Commerce of the United States Patent Office that the contents of his application "might be detrimental to the public safety of defense," and was warned to "in nowise publish or disclose the invention or any hitherto unpublished details of the disclosure of said application, but to keep the same secret."

*

Later, Pauling discussed with the OSRD the procedure for obtaining a suitable manufacturer to produce his invention. The parties ultimately decided on Arnold Beckman's company as the likely purveyor, due to its past experience with this and other highly technical laboratory equipment.

Reuben Wood was also interested in patenting certain features that he had developed, and wrote to the NDRC for permission in March 1942. Important aspects for which he claimed responsibility were a "method of balancing the test body;" an improvement "which reduces the effect of temperature changes in the indication of the meter;" and "a method of selecting range of maximum sensitivity." He later wrote to Richard Lyon enclosing four Records of Invention statements detailing his refinements of the Pauling Oxygen Meter.

However, in a letter to Captain Robert A. Lavender of the OSRD, Pauling communicated that it was not the intention of the California Institute Research Foundation to apply for patents on the inventions of J.H. Sturdivant and Reuben E. Wood. This meant that, as concerned the Oxygen Meter patent, Wood was left out in the cold.

In March 1944 the Naval Research Laboratory of Washington, D.C. sent a confidential statement to the Chief of the Bureau of Ships in which it was stated that, "this Laboratory has been interested in the development of an oxygen indicator suitable for service on submarines. The most satisfactory instrument has appeared to be the Pauling Oxygen Meter and a detailed study has been made of its operating characteristics, ruggedness, dependability and general efficiency with very promising results."

The letter also noted that the Pauling Oxygen Meter was found to be superior to a similar instrument — namely, the one created by Havens. The efficacy of Pauling's invention was becoming manifest and, as he himself had predicted, the device was proving to be of use for both the war effort and in peacetime.

Pauling's Oxygen Meter was at last patented on February 25, 1947, some five and a half years after the initial application was submitted.

Following the conclusion of the war, the meter was repurposed in particular for the incubators used to house and protect premature infants. With it, hospital staffs were better able to maintain safe oxygen levels, thus reducing the risk of brain damage and death among newborns. Pauling was proud of his instrument's peacetime applications and occasionally noted it as one of his more significant accomplishments.

Rocket Propellants

The oxygen meter was far from Pauling's sole activity in support of the war effort. Not long after the NDRC was created in the summer of 1940, Pauling was assigned to a group of scientists tasked with working on hyper-velocity guns. The committee that Pauling belonged to was specifically charged with creating a high-performance propellant to use in these weapons, which led to the creation of new experimental methods for studying powder combustion.

In 1943 Pauling began investigating a compound, dinitrodiphenylamine, that resisted the destabilization to which contemporary powders were prone. From this, Pauling's research team engineered several new powders, and his discovery led to a universal changeover from diphenylamine to dinitrodiphenylamine as a new compound that was far safer to work with in industrial settings. He filed for a patent on June 18, 1945.

The situation from there becomes a bit confused, perhaps because certain documents related to the invention may have been embargoed. Intriguingly, in a 1991 interview with biographer Thomas Hager, Pauling recalled that "I patented, during the war, a class of composite explosives — propellants. And it may be that they are used, to some extent, now. I never got any royalties from that, because the government had an irrevocable royalty-free license, and nobody else was interested in the powder for propelling bazookas and things like that."

Pieces of Pauling's claim are not verifiable, but documents held in the Pauling Papers provide further insight into what may have happened with this case. They begin on May 15, 1945, when Pauling wrote a statement in which he agreed to assign to the California Institute Research Foundation his entire right, title and interest in the "Propellant and Method of Controlling the Burning Thereof," (OEMsr 881 Pat 1) along with any patent which the Foundation might file, as long as he received a quarterly payment of 15% of the income from the invention. As Pauling recalled never receiving any royalties from the propellant, either the California Institute Research Foundation never patented Pauling's invention, or there was never any income.

We do know that Pauling was still pursuing a patent in November 1948. That month, Pauling's patent attorneys, Lyon and Lyon, wrote a letter to the Commissioner of Patents "in response to the Office Action of June 8, 1948." The letter sought to amend Pauling's patent application and included a "remarks" section in which the authors listed all of the unique aspects of Pauling's rocket propellant. According to them, "The only reference [in Pauling's amendment] which is directed to a rocket or rocket propellant, is the British reference Piestrak." (A scientist based in the U.K.) Lyon and Lyon continued, "It is inherently impossible for the propellant shown in this reference to function in the manner of applicant's propellant..." meaning that Pauling's invention was wholly novel.

Lyon and Lyon then listed all of the different ways in which Pauling's propellant was unique, noting that significant changes to burning methods and infrastructure would be required for Piestrak's approach to match Pauling's. Notably, in Piestrak's invention, one cylindrical portion of the propellant would burn completely before the next one in order to create "spaced impulses," while in Pauling's, the portions were all fast-burning.

The attorneys then compared Pauling's invention to one patented by an individual named Maxim, noting that Maxim's 1896 disclosure "consists in providing in an explosive colloid, throughout its structure, uniformly arranged cells. These cells are shown in his preferred form as being voids." The voids could also be filled with a fast-burning powder, in order to expand the flame rapidly to the walls of the cells. However, Maxim's methods did not apply to Pauling's invention because Pauling's product would be utilized in the confined space of a high-velocity gun. Likewise, Maxim's powder could only function like Pauling's on occasion and seemingly by accident. Finally, Maxim's black powder would not burn at the same rate as Pauling's compound, according to the attorneys.

The story picks back up on March 22, 1951, when Pauling wrote a memo to Lyon and Lyon titled "Patent application on explosives." In it, he compared his product to other inventions and noted that, "In our case we are interested in controlling the burning rate — in conferring upon the major propellant material a burning rate other than that characteristic of it." Pauling added that he was interested in controlling the burning rate by controlling strands, or by other special methods of manufacture of the propellant, and reported that another researcher named De Ganahl had not been able to control the burning rate of his own propellant.

On March 7, 1952, Pauling received a letter from J.P. Youtz, business manager of the California Institute Research Foundation, informing him that the application serial no. 600,043 (Pauling's rocket propellants patent), which had been pending in the Patent Office, had finally been rejected by the Examiner "in spite of the fact that there is more evidence to indicate your invention is patentable over the references cited." April 12 was the deadline for an appeal.

From there, it is not clear if Pauling pushed forward or gave up on what would seem to have been a lucrative proposition. It is possible that the process was eventually patented by Pauling and then passed on to the California Institute Research Foundation or the government. It is also possible that it was signed over to one of these entities and patented later. Or maybe it was not patented at all, and Pauling's 1991 statement emerged out of a wartime legal process clouded by secrecy.

In any case, Pauling's new method of creating rocket propellants and controlling their burning, and particularly his discovery of the stabilizing effects of dinitrodiphenylamine, is recognized today as having been an important contribution to safer working practices in the explosives manufacturing industry.

Oxypolygelatin

Another war-era project for Pauling was the creation of a blood plasma substitute called oxypolygelatin. Developed with colleagues Dan Campbell and Joseph Koepfli, this new compound seemed to be an acceptable substitute for human blood, but needed more testing to be approved by the Plasma Substitute Committee. By the time Pauling received more funding, the war was near its end and the need for artificial blood had become less pressing. As a result, oxypolygelatin never really got off the ground as an acceptable blood substitute.

As time moved forward, Pauling mostly gave up on the project, in part because widespread blood drives organized by the Red Cross and other organizations had lessened the need for his invention. But in 1946 Pauling, Campbell and Koepfli decided to file for a patent on oxypolygelatin and its manufacturing process, which they then transferred to the California Institute Research Foundation with the stipulation that one of the inventors be consulted before entering into any license agreement. They also noted that the Institute should collect reasonable royalties for the use of the invention, but only so much as was needed to protect the integrity of the invention. The "Blood Substitute and Method of Manufacture" patent was filed on December 4, 1946, and the Trustees of the Institute agreed to take on ownership of oxypolygelatin and the patent application in early 1947.

Although it would appear that Pauling largely abandoned the oxypolygelatin project with the transfer of ownership, he still pushed for its manufacture years later. In one instance, an October 1951 letter to Dr. I.S. Ravdin of the University of Pennsylvania Medical School, Pauling expressed a feeling that oxypolygelatin was not being considered seriously enough by the medical world as a blood substitute. It was Pauling's belief that "Oxypolygelatin is superior to any other plasma extender now known." He likewise noted that it was the only plasma extender to which the government possessed an irrevocable, royalty-free license, and he could not understand why it was not being stockpiled and utilized.

As far as Pauling knew, only Don Baxter, Inc., of Glendale, California, was manufacturing the substance. At this point the rights to oxypolygelatin were owned by the California Institute Research Foundation, not Pauling, and the Institute was not authorized to make a profit from it. Consequently, Pauling's insistence on the production and usage of his invention can only be explained by a concern for humanity, coupled perhaps with an urge to see the creation succeed on a grander scale.

In another 1951 letter, this time sent to Dr. E.C. Kleiderer, Pauling continued his push, arguing in this instance that oxypolygelatin was superior to the plasma substitutes periston and dextran. In Pauling's opinion, "the fate of periston and dextran in the human body is uncertain...these substances may produce serious injuries to the organs, sometime after their injection."

Oxypolygelatin, on the other hand, was rapidly hydrolyzed into the bloodstream and would not cause long-term damage. It was also a liquid at room temperature, unlike other gelatins, and was sterilized with hydrogen peroxide to kill any pyrogens (fever-inducing substances) while many other gelatin preparations failed because of pyrogenicity. One of the only problems with oxypolygelatin was that the chemical action of glyoxal and hydrogen peroxide could potentially produce undesirable by-products, but Pauling felt the matter could be cleared up with further investigation.

Pauling's interactions with Ravdin and Kleiderer did not yield much fruit, but this did not deter Pauling from pursuing the matter many years later. In 1974, after visiting Dr. Ma Hai-teh in Peking, China, Pauling forwarded a copy of his published paper on oxypolygelatin, and introduced the possibility of its manufacture in China. "I hope that you can interest the biochemists and pharmacologists in investigating Oxypolygelatin," he wrote to Ma. "I may point out that no special apparatus or equipment is needed."

Ma expressed interest in his reply and said that he had passed Pauling's paper on to a group of biochemists, before adding that he was personally more excited by Pauling's work on vitamin C. The rest of their correspondence focused primarily on the benefits of vitamin C, especially in the treatment of psoriasis.

In a 1991 interview with Thomas Hager, Pauling claimed, "I patented, with a couple of other people in the laboratory, the oxypolygelatin. I don't remember when I had the idea of making oxypolygelatin. Perhaps in 1940 or thereabouts." He added that it was never approved by the Plasma Substitute Committee because it wasn't homogenous; meaning that, on the molecular level, it included a range of weights. Pauling, however, believed that the range in molecular weights should not matter, since naturally occurring blood plasma includes serum albumin and serum globulin, whose molecular weights fall in a wide range anyway.

Regardless, because of the committee's decision, oxypolygelatin was classified as being not usable for humans but had, in fact, been produced for veterinary use. Indeed, at the time of the interview, Pauling believed that oxypolygelatin was still being manufactured somewhere in the world, but was unsure of the details since there were many rumors floating around.

In 1992 Hager also interviewed Pauling's co-inventor, Joseph Koepfli, who claimed that oxypolygelatin had indeed been deployed for human use by motorcycle officers around Los Angeles who were the first to the scene of accidents. He also remembered that, in the early 1980s, Pauling had told him that oxypolygelatin was used for years in North Korea, but that no one was ever paid any royalties.

These and a few other rumors about oxypolygelatin circulated, but evaluating their worth is virtually impossible due to the secrecy surrounding wartime scientific work, as well as the scarcity and ambiguity of the surviving documentation. Judging from Pauling's opinions though, one might suppose that, if it had been pursued more vigorously, oxypolygelatin could have benefited the war effort and proven successful on a commercial level.

Cavity-Charge Projectile

"Major Ross, patent attorney for the Navy...said that perhaps I didn't know that I was co-inventor in this invention — I do not remember having been told that I was. The invention is on an offset liner for cavity charge."

— Linus Pauling, note to self, February 8, 1952

In February 1952, Pauling was summoned by K.F. Ross, patent attorney for the Navy, to sign an oath and patent application form. The document was titled "Oath, Power of Attorney, and Petition," and stated that Pauling and Martin A. Paul were the joint inventors of "An Offset Liner for a Cavity Charge Projectile." Paul had already signed the same agreement on January 17, 1952. The document also stated that D.C. Snyder and K.W. Wonnell, attorneys affiliated with the Office of Naval Research, would manage the patent application.

When the inventors signed the patent application form, they also agreed to sell their invention to the Navy, which made the purchase for the sum of one dollar. It was furthermore stated that, "The said Owners hereby agree to execute and deliver unto the Government, upon request, any and all instruments necessary to convey to the Government the full right, title, and interest in and to any substitutions, divisions, or continuations in part of said application." In this, Pauling was able to simultaneously claim inventorship and sign away ownership, as well as any other claims to the invention, with one stroke of his pen.

The timing of the application, coupled with the absence of the cavity-charge projectile from Pauling's research notebooks, suggest that this was another of Pauling's war work projects, but one that remained top secret until after the war.

The problem that the researchers endeavored to solve was the stabilization of gun-ejected explosive shells. The contemporary method of stabilization upon which Pauling and Paul were asked to improve was to spin the shells as they were ejected, which was not very efficient. For one, spinning the shells resulted in a 50% decrease in the force that the shells could deliver upon impact, relative to a shell that does not spin. Working together, Pauling and Paul found a creative way to provide stabilization without lessening the impact that the shells could make on their targets.

The primary objective of their project was to improve the penetrating power of a spin-stabilized explosive shell by inventing an improved cavity-charge shell. A cavity-charge shell includes a space around which the explosive is arranged, so that when the explosive detonates, the shaped cavity focuses and increases the detonation, thereby requiring a smaller amount of explosive to deliver a comparable amount of force.

One tactic used by Pauling and Paul to increase efficiency was to change the shape of the cavity's liner. The new and improved model of a cavity-charge projectile utilized a plurality of offset plane sectors which faced in the direction of the shell's rotation, ostensibly causing the shell to be slowed less by spinning.

Further, in Pauling and Paul's model, the liner for a cavity-charge projectile was constructed by dividing the conical surface of the cavity into sectors, and tilting each sector slightly towards the preceding sector. According to the duo's patent, "45 degree steel cones of .062 inch thickness and sectioned in half and in quarters were respectively put together again with silver solder in such a way that adjoining edges were offset with respect to each other." Upon impact, the force exerted by the explosive in the shell on these sectors would compensate for the slowing of forward motion caused by spin.

Pauling and Paul had been constructing cavity liners by dividing a conical surface into four separate sections which were then twisted or canted relative to each other. But the patent states that a die could be constructed which would enable the structure to be made in a single stamping. As to the efficiency of the offset cavity liner, "It can be seen that for speeds of rotation above about 130 r.p.m., the modified cones were far superior to the unmodified cones."

Several variations in the invention emerged with slightly different cavity shapes and other modifications, but the patent concludes that the various versions of the invention all had key features in common. For one, each required the offset surface to face the direction of rotation of the shell. Likewise, they all required "that there be a plurality of offset sectors where the amount of offset increases from apex to the base of the shell head portion."

Pauling and Paul's joint invention, "An Offset Liner for a Cavity Charge Projectile," U.S. patent number 3,217,650, was patented on November 16, 1965, 13 years after the original application was filed.

Road Signs

Pauling's interest in patenting his ideas continued long after the conclusion of World War II, and his track record of success remained mixed, in some cases because other people beat him to the punch. Pauling's pursuit of a non-blinding road sign is an example of one such idea that seemed novel, but which had already been claimed by others.

In December 1983, Pauling wrote to James A. Thwaits, President of International Operations and Corporate Staff Services at 3M (a global innovation company responsible for inventions such as the Post-it Note) with an idea that he felt might solve an everyday problem. Pauling's concern was that he and many other motorists encountered difficulty reading road signs when the sun or a very bright sky was positioned directly behind the sign.

To solve this problem, Pauling suggested that transparent glass or plastic rods be embedded "penetrating from one side of the sign to the other." The end of the rods facing toward the sunlight would be shaped to gather light from the "sun side" and redirect it along the rods in such a manner that it outlined the words on the sign. This could be achieved with the use of "several small rods...grouped together" like fiber optics in a manner that would promote internal reflection across their surfaces. Pauling concluded that many accidents were caused by the illegibility of backlit road signs and the resultant distraction of motorists trying to make out the letters.

Later in December, Pauling received a reply informing him that his road sign idea had been forwarded to the Corporate Technical Planning and Coordination Department at 3M. In the letter, he was advised to obtain a patent for the concept in order to protect it and to aid in the idea-sharing process with 3M. If Pauling did not pursue the patent, his communication would be treated as non-confidential. He was also advised to consult a patent attorney on the patentability of his road sign. Included with the letter was a booklet titled "About Your Idea!" which discussed policies for idea submissions.

In due course, Pauling wrote to Reginald J. Suyat of the law firm Flehr, Hohbach, Test, Albritton & Herbert, outlining the non-blinding road sign idea and inquiring about its patentability. Suyat answered that, in order to ensure that Pauling's design was a novel one, the firm would need to conduct a search of the Patent Office literature, at a cost of $600. Pauling complied, and within a few weeks the firm had discovered 11 existing patents that originated from ideas similar to Pauling's, most of which had been registered between 1928 and 1939. In his response, Suyat noted,

> "The patents...disclose signs and a game which are illuminated by reflective sunlight or artificial lighting. Light is transmitted through translucent or transparent inserts. In particular, Slutsky...discloses the idea of a sign whereby sunlight is transmitted through openings formed in the sign to cause sign characters to be visible from the front side of the sign. Nelson, *et al.* ...also disclose that a sign may be placed such that sunlight from the rear of the sign would be transmitted through translucent members in the sign. Speers...discloses light-transmitting pegs, while Gill...discloses a translucent member with opaque material applied thereto."

Thus presented with convincing evidence that his idea was already taken, Pauling abandoned his road sign and directed his energies elsewhere.

Superconductors

"I believe that a discovery that I have made may make room-temperature superconductivity a reality."

— Linus Pauling, June 1990

In the late 1980s, Pauling became interested in a new and exciting scientific endeavor: high-temperature superconductivity. While most of the field's researchers at that time were investigating the use of ceramics to promote superconductivity, Pauling decided to focus more on techniques for raising the temperature at which materials became superconducting. Doing so, he reasoned, would facilitate their usage in industrial and research settings. High-temperature superconductivity, or high-Tc, was a technique discovered in 1986, so in early 1988, when Pauling took up the topic, the field was still wide open to exploration.

But what is high-temperature superconductivity? According to a 1988 business agreement drawn up between Pauling and IBM, the definition of a "superconducting product" is "any product which contains any material which loses substantially all electrical resistance below a transition temperature above 77 degrees Kelvin." In other words, a superconductor is a substance that loses electrical resistance when heated to a point between 77 degrees Kelvin and some higher temperature.

(It is important to note that the high temperatures being discussed in the context of superconductivity are actually quite cold: 77 degrees Kelvin translates to –196 degrees Celsius. Superconductivity has traditionally been observed at temperatures near absolute zero; achieving it at something near room temperature would constitute a major scientific breakthrough.)

Until the late 1980s, the generally accepted theory of electric superconductivity of metals was based on an understanding of the interaction between conduction electrons and electrons in crystals. The critical temperature of superconductivity was thought to be below 23 degrees

Kelvin (roughly –418 degrees Fahrenheit), but in the late 1980s, it was discovered that superconductors could have critical temperatures above 100 degrees K, which threw the theoretical understanding of the subject into confusion and controversy. The discovery also spurred an effort to find new materials with an even higher Tc, or temperature of superconductivity; perhaps as high as room temperature.

Pauling's first step in exploring high-temperature superconductors was to contact Zelek Herman, a biochemistry professor at Stanford and close colleague at the Linus Pauling Institute of Science and Medicine (LPISM). Pauling's somewhat unusual request was that Herman create a few color slides for him of the cover of *American Scientist*. The particular issue in which he was interested depicted the structure of a high-temperature superconductor.

A month later, Pauling wrote to Herman again, this time about the possibility of obtaining a Naval Research grant to fund an investigation of the "resonating valence bond theory of superconductivity." In developing the proposal, Pauling emphasized the importance of both fluxon theory and a method of calculating interaction with phonons by using the relation between bond length and bond number. The latter method had been formulated by Pauling in 1947.

As he detailed to Herman, an idea for creating a superconductor had occurred to him while he was thinking one day about how Damascus steel was made for swords in the Middle Ages. The exact process by which Damascus steel was originally fabricated is unknown, but one way of reproducing it is through billet welding, where layers of steel are folded over and over and then stretched until a desired thickness is reached.

In February 1988, Pauling decided to apply this method to the building of a superconductor, using lead and a malleable plastic. The idea was to see if he and his associates could get the lead thin enough to become superconducting. Pauling named his idea, "A method of fabricating a composite containing filaments of a superconducting material with diameter and cross-sectional shape such as to confer on the material improved properties, such as increase in the characteristic superconducting temperature."

Pauling's proposed superconductor leaned on the idea of phonon dampening, a process that consisted of taking a conductive metal such as tin, drawing it to a very fine diameter — specifically 10–20 angstroms (one angstrom is equivalent to one-ten billionth of a meter) — then insulating the metal with non-conductive material, such as glass. Doing so would raise the superconductive temperature, or Tc, of the metal. Pauling worked on the project together with LPISM associates at a facility that the Institute leased at the Stanford Industrial Park in Palo Alto, California.

As the work progressed, Herman came up with a creative approach for collecting material for the superconductor. His method called for inverting a bicycle, taking the tire off of one wheel, setting the wheel on a block of wood, heating a tin fiber above a furnace with torches, turning the wheel using the bicycle's pedals, and collecting a thin strand of material on the rim of the wheel. Pauling was very engaged in the process and would occasionally drop by to assist in the experimentation, sometimes by wielding the torch used in stretching the borosilicated tin while standing over an 800-degree furnace.

Many pages in Pauling's research notebook from that time show that he was likewise researching and working on calculations related to superconductors. The calculations first start to appear in February 1988 and, by Spring, he believed he had enough material to patent his idea. He filed an application for his "Technique for Increasing the Critical Temperature of Superconducting Materials" on May 31, 1988.

*

The process of developing a potentially viable superconducting product took several months and much collaboration. Along with Zelek Herman, other members of LPISM including Emile Zuckerkandl, Ewan Cameron and Steve Lawson worked on the project. When the researchers finished the task, Pauling was ecstatic and invited Herman and Lawson to his home, giving them large mineral crystals as gifts and offering to inscribe their copies of *General Chemistry* to commemorate the occasion.

The group's invention aimed to form a composite structure in which superconducting materials assumed the form of fine strands embedded in a wave-guiding matrix. The matrix restricted the superconducting current to a linear motion; however, the strands did not need to be straight, but could also be bent or interconnected into a network. This matrix would be built of a non-conducting material such as glass.

Once the superconducting material was mounted with the help of the matrix material, the entire set-up was stretched to minimize the diameter of the superconductive strands, a process that maximized the critical temperature. Optimum strand diameters were thought to lie in the range of 50–2,000 angstroms. For its part, the matrix material needed to be easily drawn into fine strands and not be superconducting. Pauling believed that "by selecting the best superconducting and matrix materials and the optimum strand diameter, it should be possible to obtain a composite superconductor with critical temperature above room temperature, critical magnetic field above 100 tesla, and critical current density above 108 amperes per square centimeter."

In the group's patent description, a few variations on this technique were listed that were thought to increase its effectiveness. One variation involved the embedding of two types of superconducting materials into the matrix instead of one. A suitable composite structure of this type could include strands of lanthanum and tin embedded in glass with a softening temperature of about 950°C.

The description also noted a couple of different ways that the matrix material and superconducting material could be joined together. In one variation, the matrix was constructed as a tube and the superconducting material poured in and afterwards "drawn," or stretched. Then several of these tubes containing superconducting material were joined together and stretched simultaneously, over and over, the same way Italian millefiori glass beads are made. Another variation utilized the filling of a porous matrix with a liquefied superconductor, whereupon the whole apparatus was heated and stretched.

The group admitted to problems with these methods, but Pauling also had solutions in mind. One obstacle was that the melting point of glass might be lower than that of the superconducting material, which would make it impractical to draw glass or other material with the superconductor. Pauling's corrective for this problem was to add a powder made up of the superconducting material to the glass in order to reinforce it.

Despite all the effort that Pauling and many other scientists were expending, a *New York Times* article published on October 16, 1988 declared that the U.S. was falling behind Japan in the race to commercialize superconductors. The piece's author offered that while "major uses of the new materials are considered to be at least ten years away", many "scientists envision superconductors that could eventually be used to make computers that operate at blazing speeds, highly efficient electric generators and transmission lines, and high-speed trains that would be suspended above their tracks by superconducting magnets."

The article continued that the new superconductors could conduct electricity at temperatures as high as −235 degrees Fahrenheit, whereas previously it had been thought that superconductivity could occur only at about −420 degrees Fahrenheit. The new temperature, the article concluded, would be much easier to achieve in laboratories.

Richard Hicks, vice president of LPISM at the time, wanted to license Pauling's invention, "Technique for Increasing the Critical Temperature of Superconducting Materials," to U.S. companies, but was met with little positive feedback. Undeterred, he instead attempted to license the invention to Japanese companies after hearing that Japan was also interested in the commercialization of superconductors. No Japanese companies showed interest either, but the Central Intelligence Agency (CIA) did come calling to ask why the Institute wanted to license a patent out of country. Over the course of their interview, the CIA representative demonstrated extensive knowledge of and interest in the project. In explaining the Institute's position, Steve Lawson clarified that no American companies had been keen to support the work, so LPISM was compelled to look to other countries.

In 1988, the same year that the LPISM research group had begun work on the high-temperature superconductor, Pauling, Hicks and Zuckerkandl set up the Superbio Corporation to administer the business side of the invention; Pauling assumed the role of chairman and Richard Hicks was president. Pauling believed it would be successful and invested in the company, owning 300,000 shares in Superbio, Inc. by the end of August. On August 12, 1988, Superbio entered into discussions with the Du Pont Company, which wanted to evaluate Superbio's information on superconductivity with a view to "possible business activity." In turn, Du Pont Co. was sworn to secrecy regarding the company's research.

Not long after, on August 31, Pauling and IBM drew up a draft agreement in which IBM agreed to purchase the patents and/or patent applications for high-temperature superconductivity from Pauling for the sum of $10,000. The document described Pauling's invention in detail, stating that it "provides a technique for increasing the critical temperature, critical magnetic field, and maximum current density" of superconducting materials. In addition, IBM was to pay Pauling "a royalty of five percent of the manufacturing cost of the patented portion of any apparatus made." The patent would become fully paid when IBM had compensated Pauling to the tune of $2 million.

In early 1989, *Superconductor News* affirmed the fears voiced by the *New York Times* the previous fall that the U.S. was falling behind Japan in the race to commercialize superconductors. The publication's January/February issue included a report on presentations given by the United States Superconductor Applications Association (the SCAA), which included Japanese developments in "SC power transmission, SC magnetic energy storage, SC generators, SC electromagnetic ships, SC electronics and computers, and the SC linear motor car (maglev)." *Superconductor News* also discussed the possibility of impending confirmation of superconductive materials that could operate at room temperature (Ambient Temperature Superconductors, or ASCs). Potential uses for room temperature avionics applications were listed as thermoelectricity, solid state synchron sources for X-ray lithography, and applications for earth and planetary sciences, medicine, biology, and physical sciences with Extra Low Frequency (ELF) magnetometry.

In response, the Exploratory Research and Development Center in Los Alamos, New Mexico was set up to boost the U.S.'s superconductivity research infrastructure. The Center expressed interest in collaborating with Pauling after he sent them a letter in July 1989 that mentioned his patent application, which by that point had been turned over to Superbio. Pauling's faith in the company was evident: by the end of November 1990, he owned 900,000 shares of common stock. Bolstered by the seeming momentum of the start-up, the interest of other companies in Pauling's superconductivity invention, and a patent in the making, the future for this work looked promising.

*

As they continued to explore possibilities for creating improved superconductors, one issue that the Pauling group confronted was verifying that their technique was successful. This was in part due to the fact that measuring changes in superconductivity was very difficult, given the small diameter of the tin used and the challenges faced in gaining ample contact.

Attempting to gauge the efficacy of their superconducting material, Zelek Herman sent a sample to Howard Hart at the General Electric Research and Development Center in Schenectady, New York, requesting that Hart run tests of their samples to determine a temperature of superconductivity, if one existed at all. It is from this communication that we are able to learn much about the samples that the group created.

Herman sent two vials to be tested, with a third vial to be used as a reference. The first vial contained a segment of a superconducting niobium-titanium seven-stranded cable about 9 cm long. This was produced by snipping off a piece of one of the strands, immersing it in

90% formic acid for one hour, rinsing it with deionized water, and drying it with a paper towel to remove residual heavy formvar. Then it was rubbed with fine steel wool and blown with compressed air. The contents of this first vial were to serve as the reference for testing the wires in vials two and three.

The second vial contained a piece of similar length, but the mass was much less and it was 0.05 mm thick, as opposed to the reference sample's 0.8 mm diameter. The third vial contained three pieces of square wire ranging between 9 and 11 cm, each of which had a diameter of about 0.3 mm, and estimated to be about 4.8% superconducting material. At the end of Herman's accompanying letter to Hart, he added that he did "not expect any elevation in Tc for any of these samples; rather, this is a test to see if you can detect a superconducting transition for samples containing a small, but continuous, amount of superconductor." Hart's reply, if one was sent, is not held with the copy of Herman's original letter in the Pauling Papers, and the results of his tests are unknown.

*

First drafted in May 1988, a copy of the patent application for Pauling's "Technique for Increasing the Critical Temperature of Superconducting Materials" was returned to Pauling on December 7, 1990, with requests from the examiner to make some changes. The deadline to return the application was tight but Pauling complied, submitting his revisions on December 12.

Finally, on October 27, 1992, the group's four years of work came to fruition in the form of a patent. The "Method of Drawing Dissolved Superconductor," Patent No. 5,158,588, was a continuation of application serial number 7/366,574, which was filed on June 15, 1989. The June 1989 application was, in turn, a continuation of the initial patent application number 7/200,994, filed May 31, 1988.

In the abstract for the "Method of Drawing Dissolved Superconductor," it is stated that "a preform for drawing superconducting wire is prepared by mixing fine particles of a superconducting material, containing barium, potassium, bismuth and oxygen, with a solvent, containing potassium hydroxide, in a tube." After the preform is heated and drawn, the superconductive material dissolves in the solvent and deposits from the solution as "a solid network of crystals in contact with one another."

The final patent included essentially all of the components of the previous patent applications. The invention provided "a technique for increasing the critical temperature, critical magnetic field, and maximum current density of any of a range of already known superconducting materials." As in descriptions of the invention submitted with previous patent applications, the structure was the same: superconducting material in the form of fine strands was embedded in a "wave-guiding matrix" which was to be made of some non-conductive material.

In its "Overview" section, the patent states that the superconducting strands within the matrix are generally round and that optimum strand diameters should lie in the range of 50–2,000 angstroms. Various techniques are provided: using one method, a composite billet could be formed of bars of the superconducting material surrounded by bars of the matrix material and then stretched until the desired diameter was reached.

A different method entailed the use of a porous matrix, such as an artificial zeolite or an aluminosilicate, the pores of which are filled with the superconducting material. This done, the entire ingot could be stretched to reduce the diameter of and align the superconducting strands. Another aspect of the invention proposed that strands of a crest superconductor and strands of a trough superconductor could be "alternating and insulated from one another in the matrix." The relative amounts of the two superconductors would minimize phonon interaction.

"Method of Drawing Dissolved Superconductor" was one of the last inventions that Pauling patented and among the last lines of research that he pursued after a lifetime of scientific accomplishment. Collaborator Steve Lawson noted in an August 2011 interview that Pauling's goal in pursuing the superconductor patent was to raise money for LPISM.

By the time that his superconductor invention was patented, Pauling was in his early 90s and not interested in adding to his personal wealth; rather, he hoped instead to help stabilize the Institute's chronically shaky finances (see Chapter 20).

After Pauling's death in August 1994, the project fell into neglect, primarily because stable funding could not be secured. However, super-conductors continue to be important today in a wide range of uses, including Magnetic Resonance Imaging machines, maglev ("magnetic levitation") trains and electric generators.

Cold Fusion

At a press conference in Salt Lake City held on March 23, 1989, electro-chemists Martin Fleischmann of the University of Southampton (U.K.) and Stanley Pons of the University of Utah made the blockbuster claim that they had achieved nuclear fusion at room temperature in a labo-ratory in Utah. If true, the discovery had the potential to revolutionize energy science and conceivably change the socio-economic fabric of the entire world.

The announcement came in the wake of a series of experiments in which Fleischmann and Pons had attempted to enable fusion by forc-ing deuterium ions into a palladium cathode using electrolysis. During their electrolysis process, an electric current was passed through "heavy water" — water that contains the hydrogen isotope deuterium — and split the water into its constituents of oxygen and deuterium.

Fleischmann and Pons's big breakthrough occurred while the duo were carrying out some exploratory tests. In the midst of these tests, a 1-cubic-centimeter block of palladium disappeared in an explosion that occurred overnight. The explosion, nuclear or otherwise, also demol-ished part of the building where the experiments were taking place. Fleischmann and Pons were motivated by this event, destructive though it was, to further pursue what appeared to be cold fusion. From then on, they kept a careful account of the power output and input of their experiments.

After a few weeks of subjecting the heavy water to, at first, .05 amps, then .1 amps, and finally .2 amps of electricity, Fleischmann and Pons recorded an excess heat output of about 25%. Heat output is an indicator of nuclear fusion, but the duo could not find evidence of neutron production, another indicator of fusion. However, learning that Steven E. Jones of Brigham Young University, who had worked on muon-catalyzed fusion, had observed weak evidence of neutron production from cold fusion experiments, Fleischmann and Pons were encouraged to believe that their own experiments were probably producing neutrons as well.

Their morale boosted by this bit of news, and feeling some measure of institutional pressure to spread the word of what they may have uncovered, the scientists published their findings and participated in the March 23 press conference. The scientific community immediately began to scrutinize their published data, keen on either confirming or debunking the phenomenon of cold fusion. But the reviewers met with mixed results: no one could reproduce the required results of excess heat and neutrons, perhaps because many were still uninformed on the exact details of Fleischmann and Pons's experiments. Meanwhile, the media speculated that this new form of energy could be the answer to global concerns over diminishing fuel supplies, sparking international furor about cold fusion and producing varying accounts of the original experiments.

A month after the press conference, Pauling wrote a letter titled "Explanations of Cold Fusion" to the editor of *Nature*. In it, Pauling noted that palladium is saturated with hydrogen at the composition $PdH_{0.6}$. This given, Pauling suggested that the introduction of additional hydrogen atoms brought about by the Fleischmann-Pons experiments had caused extra deuterons to be forced into the palladium cathode to form an unstable higher deuteride, PdD_2. This instability resulted from the free energy of the EMF (Electromotive Force) used during electrolysis, and also because palladium is saturated with hydrogen at the composition $PdH_{0.6}$. According to Pauling, it was the decomposition of this unstable deuteride that caused the increased heat observed by the scientists. In other words, what Fleischmann and Pons observed was not an occurrence of cold fusion.

Pauling further opined that the unstable higher deuteride PdD_2 "may begin to decompose either slowly, resulting in the increased liberation of heat, or explosively, as was observed when a 1-cm cube of the deuterated palladium disappeared" overnight in Fleischmann and Pons's laboratory. Pauling believed that "because of the difference in amplitude of the zero-point vibrations of the nuclei with different masses, palladium dihydride would be less stable than palladium dideuteride." Reasoning that the decomposition of the unstable compound was causing energy output to exceed input, Pauling provided the world with a rational explanation for why cold fusion was not occurring.

*

Pauling's letter was published in May 1989, but it did not mark the end of his interest in the subject of cold fusion. In May 1992, while at home on his ranch in Big Sur, California, Pauling had a conversation with his grandson Barclay J. "Barky" Kamb, during which he revealed his idea for a nuclear fusion invention.

Pauling had taken note of the fact that many experiments reported a "liberation of neutrons or helions or other indication of nuclear reaction greater than the background count," but that not all interested researchers had observed the phenomenon. As he thought about the problem, he reflected back on his 1989 letter to *Nature* in which he suggested that the decomposition of small amounts of PdH_x were responsible for thermal anomalies, and that related explosions, including one that killed an SRI International researcher, were due to large amounts of PdH_x decomposing.

Branching off from this train of thought, Pauling had an idea for increasing the amount of energy within certain compounds, and came up with a few theories on how to maximize the amount of energy held by particles in order to achieve cold fusion. He hypothesized that "the stored energy in PdD_x, $x > 0.6$, might be produced either by high pressure of $H_2(D_2T_2)$, (heavier hydrogen isotopes) with Pd, Ti, or other metals." The metastable or unstable compounds resulting from this high-pressure compound could then be heated with the use of converging laser beams or through the application of thermal energy, which could be obtained by chemical explosives.

From there, Pauling suggested utilizing the Monroe Effect in conjunction with the chemical explosives. The Monroe Effect arises when one cuts a hollow into the surface of an explosive with the intent of focusing the force of a blast. When combining the surface cut with a conical liner of PdD_x, or a similar unstable metal like deuteride, the technique works to direct a blast toward a particular location.

As an alternative, Pauling also proposed superheating pellets made of compounds such as $M(H, D, T)_x$ and using laser beams or explosions to increase the energy within the compound. In essence, Pauling's idea was to increase the energy stored in $Pd(H,D,T)_x$, $x > 0.6$, or $M(H,D,T,X)_x$, $M = PdTi$, by applying external sources of energy in specific ways to catalyze fusion.

Pauling speculated that an augmented detonation could produce shock waves that would accelerate particles, perhaps along channels in the metals, to prompt fusion by reaction. His proposed methods of increasing stored energy involving $M(H,D,T,X)_x$ included shooting pellets of the compound into a heated chamber, utilizing plasma in a tokamak (a donut-shaped device used in hot fusion which uses a magnetic field to confine a plasma) or focusing a detonation wave within conical metal or something similar.

Clearly there were many pieces to Pauling's invention claim, but they all described methods for increasing the yield of nuclear fusion energy. All of the methods also depended on an increase in the momentum of the atomic nuclei involved, an increase provided by a source of energy supplementary to the stored-up energy of a given high-energy compound.

A month after Pauling's conversation with Barky, Pauling followed up with a letter to his grandson in which he detailed two additional ideas. The first was to augment the internal energy of portions of PdD_x, or other high-energy materials, by introducing portions into a rotating cylinder containing "spheres or other aggregates of hard materials, such as steel or other hard metallic alloy...such as to cause vigorous contacts of these spheres or other aggregates with one another." The object of these collisions, again, was to add to the internal energy of the materials.

Another similar idea for augmenting the yield of nuclear fusion energy was, Pauling suggested, "by a method, similar to a ball mill in the manufacture of Portland cement, in which there is a rotating cylinder containing spheres or other aggregates of hard materials that can collide with one another..." Portions of "palladium or titanium or other alloy with deuterium or tritium or other fusionable nuclei" would then be introduced into the mix, producing high-energy material. Pauling felt that the excess heat emerging from reactions of this type could be utilized for generating electric power, and that the unreacted alloys could be reused as additional spheres or aggregates.

Although Pauling tinkered around with these methods of prompting fusion with the notion of someday patenting them, the ideas lay fallow and, a little over two years later, Pauling passed away. His notes, however, remain useful insofar as they contribute to the ongoing conversation about the possibility of cold fusion and of ways of facilitating hot fusion. Pauling's thoughts on modern subjects such as nuclear fusion and cold fusion were also further evidence of an active and inquisitive mind even as he neared the end of his life.

Lipoprotein(a)

In the late 1980s and into early 1990, Pauling and a colleague, Matthias Rath, worked intensively on the health benefits of vitamin C and Lipoprotein(a) binding inhibitors. In 1990 they applied for two patents related to that research. The first, applied for in April, was titled "Use of ascorbate and tranexamic acid solution for organ and blood vessel treatment prior to transplantation." The second, submitted in July, was titled "Prevention and treatment of occlusive cardiovascular disease with ascorbate and substances that inhibit the binding of lipoprotein (A)."

The technique that Pauling and Rath were attempting to patent in April was both a method and a pharmaceutical agent designed to 1) prevent and treat fatty plaque buildup in arteries and organs, and 2) prevent blood loss during surgery by introducing into a patient (or organ) a mixture of ascorbate and lipoprotein(a) [Lp(a)] binding inhibitors, such as tranexamic acid.

Tranexamic acid is a synthetic version of Lysine, and ascorbate is the shortened name for L-ascorbic acid or, more commonly, vitamin C. Lp(a) is a biochemical compound of lipids and proteins that binds to fibrin and fibrogen in the walls of arteries and other organs. This binding causes plaque buildup, which in turn often results in atherosclerosis — the thickening and embrittling of arterial walls — and cardiovascular disease (CVD), one of the most common causes of death in the United States. The second patent described effectively the same method, but focused more on CVD and less on surgery.

In their research, Pauling and Rath pointed out that humans and a select few other animals are the only creatures that suffer from heart attacks and other issues associated with the buildup of plaque in the circulatory system. One common link between all of these creatures is the fact that they do not naturally produce vitamin C, and therefore must obtain it solely through diet. The duo hypothesized that the cause of Lp(a) buildup was a chronic lack of vitamin C, and that if vitamin C intake was increased, it would help the body filter out Lp(a) and therefore decrease the amount of Lp(a) in the bloodstream. To test their idea, the duo ran trials on Hartley guinea pigs, since they are one of the few other animals that do not synthesize their own vitamin C.

The first test was run on three female guinea pigs, each about a year old and weighing 800 grams. The animals were all fed a diet devoid of ascorbate (a hypoascorbate diet) but given a daily injection of ascorbate, such that Pauling and Rath could easily monitor and control their intake. The first pig was given ascorbate at a ratio equivalent to 1 mg per kilogram of body weight (1 mg/kg BW). The second pig was given 4 mg/kg BW, and the third was given 40 mg/kg BW.

The experiment only lasted three weeks, because Pauling and Rath didn't want to guarantee a painful demise for the guinea pigs. Creatures deprived of vitamin C for prolonged periods develop scurvy, an agonizing condition where the victim becomes lethargic and begins to suffer skin color and texture changes, easy bruising, brittle and painful bones, poor wound healing, neuropathy, fever, and eventually death.

The guinea pigs had their blood drawn at the start of the test, then once again after ten days. At the end of three weeks, the animals were anesthetized and euthanized, then dissected. This investigation showed that the hypoascorbate guinea pigs had noticeably higher plaque buildup and general amounts of Lp(a) in their bloodstream. Upon closer analysis of the organs and the arterial wall, the researchers discovered that the guinea pigs had also developed lesions along the walls of their arteries, to which Lp(a) was binding even more than normal.

Pauling and Rath then ran a more expansive second trial, with a test time of seven weeks and a test group of 33 male Hartley guinea pigs, each approximately five months old and weighing 550 g. At the outset, the subjects were split into multiple groups. Group A consisted of eight guinea pigs and was given 40 mg/kg BW of ascorbate daily, while Group B consisted of 16 guinea pigs given 2 mg/kg BW daily. At five weeks all of Group A was euthanized and studied, as was half of Group B. The second half of Group B then had their daily dosage increased to 1.3 g/kg BW for two weeks before being euthanized.

Once again, it was observed that the hypoascorbate guinea pigs had developed lesions in their arterial walls and organs, as well as increased plaque buildup and Lp(a) levels. On the flip side, the second half of Group B showed decreased levels of Lp(a) in their blood and decreased amounts of plaque after their ascorbate intake was dramatically increased.

Pauling and Rath felt that their research was confirming their hypothesis, and wanted to see how it would function on humans. Their method here was to obtain post-mortem pieces of human arterial wall. They then cut the pieces into smaller sections, and for one minute placed a piece weighing 100 mg into a glass potter containing 2.5 mL of a mixture of ascorbate and tranexamic acid. Compared to the other pieces, the portions in the mixture released sizable amounts of Lp(a).

*

With these promising results in hand, Pauling and Rath decided to create a patentable treatment. Their approach rested on three main principles: First, that increased vitamin C levels in the bloodstream would prevent the creation of lesions to which Lp(a) might bind. Second, that lipoprotein binding inhibitors would detach any plaque that had already built up. And third, that vitamin C would then also help the body to filter out Lp(a). In this way, it could be used to both treat and prevent CVD and other related cardiovascular problems.

The duo also saw great potential use for their research in surgery — specifically angiopathy, bypass surgery, organ transplantation, and hemodialysis. Lysine or other similar chemicals naturally help to speed up the healing process and also act as blood-clotting agents, thereby reducing the risk of blood loss during surgery. Also, patients undergoing organ transplant surgery, bypass surgery, and hemodialysis often suffer strong recurrences of CVD, which Pauling and Rath felt was due to depleted vitamin C levels from blood loss. Similarly, diabetics often struggle with inhibited vitamin C absorption and higher levels of Lp(a), leading Pauling and Rath to hope that their work could help to treat diabetes-related CVD as well.

When living patients were using their treatment, the mixture was designed to be taken orally in pill or liquid form, or injected intravenously. Pauling also wondered if the mixture could be taken subcutaneously (injected into the deepest level of skin), percutaneously (injected into internal organs), or intramuscularly (injected into the muscle). When being used as preparation for transplant surgery, the organs to be transplanted were to be soaked in the mixture.

(Later research done by other scientists showed that there are specific vitamin C carrier molecules in the digestive tract that limit the amount of vitamin C a person can absorb when taken orally. As such, injection is a much more effective method of getting vitamin C into the bloodstream.)

Pauling and Rath's work was polarizing, if not unprecedented. As far back as the early 1970s, enthusiastic support for vitamin C by Pauling and others had been a point of extreme controversy. Now, even with this latest batch of research, many scientists and doctors seemed to think that their conclusions were grossly overstated, and in some cases even dangerous. For their part, Pauling, Rath, and their supporters felt that the harsh criticism emerged, at least in part, from pharmaceutical companies concerned about losing revenue if people stopped buying their expensive medications and instead bought inexpensive, common vitamin C.

Though the controversy wasn't new, it came as a disappointment for Pauling. He repeatedly expressed these feelings, pointing out that he was not the first to make claims about the benefits of vitamin C nor even the most extreme, and yet he was viewed as a controversial figure espousing fringe medicine. He also noted that the Pauling-Rath team was not the only group researching and promoting the positive effects of vitamin C. Other labs, such as that led by Valentin Fuster of Harvard Medical School, were conducting similar experiments. Pauling and Rath attempted to collaborate with them where possible, often with success. But more generally the duo had to rely upon individual case histories to support their research, largely because they were unable to convince major American institutions to conduct their own studies or to sponsor LPISM in its pursuits.

*

On July 27, 1993, Pauling and Rath were awarded a patent for the application filed in April 1990. On January 11, 1994, they received a second patent for the application filed in July 1990. Shortly afterward, in March 1994, the two filed a third application, following similar grounds, titled "Therapeutic Lysine Salt Composition and Method of Use." The compound they were patenting was a mixture of ascorbate, nicotinic acid (also known as vitamin B3 or niacin) and lysine, or a lysine derivative. The mixture was to be combined at a ratio of 4:1:1, and include a minimum of 400 mg of ascorbate, 100 mg niacin and 100 mg lysine. The mixture functioned more or less identically to the previous two patents, the major difference being the inclusion of vitamin B3 for its antioxidant properties. Pauling and Rath also encouraged the inclusion of additional antioxidant vitamins.

The third patent application was approved and awarded to Pauling (posthumously) and Rath in 1997. The two hadn't made any profit off the previous patents to speak of, and research that followed in the later 1990s and after 2000 put a damper on their claims. The discrepancy between the Pauling-Rath trials and subsequent tests seems to be attributable to the major differences between the two test subjects — humans and guinea pigs. However, other trials have shown that large doses of vitamin C are useful in fighting CVD — for reasons other than Lp(a) levels — and also work to combat stroke, decrease blood pressure, and provide other health benefits.

Two More from Pauling and Rath

In 1991 Pauling and Rath drafted two patent documents not related to their Lp(a) research, documents that did not ultimately result in finalized patents. One described an attempt to use synthetic polypeptides to prevent disease by helping to synthesize an optimum level and strength of collagen in the body.

"Polypeptide and Methods of Use," application drafted July 10, 1991:

Polypeptides are linear chains of two or more amino acids linked by a covalent bond. Many scientists had asserted that synthetic polypeptides would be ineffective because polypeptides are fairly conservative molecules and, as such, trying to recreate them would result in substances with little or no potency. Pauling disagreed with this sentiment completely and utilized synthetics for the purpose of his research because they were fairly easy to manufacture.

Pauling and Rath believed that synthetic polypeptides would remain viable and that arguments against them were based on a fundamental misunderstanding of what makes a polypeptide potent in the first place. A polypeptide chain with an arginine-glycine-aspartic acid (RGD) sequence was, specifically, what the duo was investigating, and the RGD sequence is the piece of the puzzle that many scientists felt would be a source of potency trouble in synthetics.

Pauling and Rath argued that the RGD sequence was not actually important, but that the R and D *was* important. Specifically, beginning a chain with R (arginine) then ending it with D (aspartic acid) — both highly polarized end peptides — was the key to imbuing a polypeptide with strength and utility. In the eyes of the two researchers, if R and D were in the right spots, it did not particularly matter what resided in between them.

Pauling and Rath's polypeptide treatment was designed to treat diseases that were related to cell migration or cell membrane adhesion. Because the treatment would cause certain cells more difficulty in penetrating membranes (or migrating in general), it would ideally contain diseases such as metastatic cancer. Also, by preventing membrane penetration and adherence, diseases such as infectious viral agents — which rely on doing just that to spread — would be contained as well.

"Treatment of Pathological States Related to Degeneration of Extracellular Matrix System Treatment," application declaration drafted November 1, 1991:

The extracellular matrix (ECM) provides structural support to animal cells, is the defining feature of connective tissues, and serves other important duties in the cellular structure. Pauling and Rath's November patent idea was for, once again, a vitamin C mixture, this time designed to prevent the deterioration of the ECM, thought to both contribute to and be characteristic of the spread of diseases, specifically metastatic cancer.

In this instance, Pauling and Rath's research revolved around apoprotein(a) [apo(a)] which, they theorized, acted as a sort of temporary surrogate to vitamin C. When vitamin C levels in the bloodstream drop, both apo(a) and Lp(a) levels increase. Apo(a), a crucial component of the body's defense against disease, was seen to be acting as "temporary vitamin C," which in the short term was beneficial, but after longer periods of time would actually contribute to ECM deterioration and other health issues.

Pauling and Rath worked with LPISM colleagues Aleksandra Niedzwiecki and Jerzy Jurka on this project, and they all concluded that apo(a) helped the body to fight free radicals and other diseases. That said, the team also felt that apo(a) needed the help of large amounts of vitamin C to be effective, especially because, as the body became sick or fought off illness, vitamin C levels in the blood dropped. As such, large doses of vitamin C were the best course of action to ensure the strength of the ECM and subsequent general health.

Matthias Rath departed from LPISM in 1992 and Aleksandra Niedzwiecki moved to work with him at his new institute. As noted, Linus Pauling was already fighting his own cancer at that time and ultimately died in August 1994. As a result of both events, the patent applications initiated in support of the two lines of research described above appear not to have been vigorously pursued.

19 Jack Kevorkian and a Twist on Capital Punishment

By Chris Petersen

Orig. Paper   AAAS — Dec. 1958

~~Reprinted from CRIME IN AMERICA~~
~~edited by Herbert A. Bloch~~
~~published by Philosophical Library~~
~~15 East 40 Street, New York 16, N. Y.~~

Chapter 7

CAPITAL PUNISHMENT OR CAPITAL GAIN?

JACK KEVORKIAN, M.D.

Title page of Kevorkian's 1958 paper on capital punishment and medical research.

Jack Kevorkian, the medical pathologist famed for his active support and engagement with physician-assisted suicide, achieved international notoriety — and eventually imprisonment — for assisting with the premature deaths of terminally ill individuals. But well before he became widely known for this work, Kevorkian lobbied in favor of a different death-related practice, one that came across Linus Pauling's desk in the early 1970s.

As first outlined in a 1958 paper, Kevorkian's proposal was that condemned prisoners be given the option of essentially donating their deaths to science. The crux of Kevorkian's idea, which was delivered at a meeting of the American Association for the Advancement of Science, went as follows:

> "I propose that prisoners condemned to death under capital punishment be allowed to submit, by their own choice, to medical experimentation under surgical anesthesia, to be induced at the set minute of execution, as a form of execution in lieu of the conventional methods prescribed by law. [...] Most of us are well aware that the ultimate 'laboratory' for testing every medical fact, concept or device is man himself. [...] In this logical and proper sequence of trial, the human subject, at the end, the 'guinea pig,' is, and always will be, the most difficult link to procure. [...] Viewing the problem purely realistically, capital punishment, as it exists today, offers an unrivaled opportunity to break these limits. It can do this by introducing into the situation an involuntary factor without destroying the necessary safeguard of consent."

Kevorkian then suggested that these types of medical experiments

> "should be funneled from all over the civilized world into a central agency, perhaps under the auspices of the United Nations. [...] Of necessity, the experiments should be extremely imaginative, should deal with things completely uninvestigatable in living men under usual circumstances. [...] Multiple experiments may be done simultaneously or sequentially on different parts or systems of a single body and could conceivably last for days under uninterrupted anesthesia stringently controlled by at least two board anesthesiologists. If experimentation does not cause death, it would ultimately be induced by an overdose of the anesthetic agent."

Kevorkian recognized that there would likely be "real and valid criticisms" to his proposal "from the legal side," and that "for society, there would be some added financial expense." But for the convicted, Kevorkian saw few disadvantages and "for medical research, there is no seeming disadvantage."

Indeed, from the perspective of the prisoner, the prospect of dying for science could even mitigate a major philosophical objection to capital punishment. Namely, in the case of the execution of innocent persons (presumed to be a remote possibility), it would give the victim perhaps a little solace to know that an ugly act of human injustice might still provide some benefit to humanity writ large.

Kevorkian would extend these thoughts in later articles and compile many of them in a 1960 book, *Medical Research and the Death Penalty*. He also actively solicited the support of notable figures in favor of his idea. It was in this context that, in January 1973, Kevorkian sent Pauling a packet of materials that included copies of letters that he had received from a number of scientists and physicians expressing their support — sometimes hedging — of the proposal.

Pauling appears to have lent a token of approval via a phone conversation between he and Kevorkian not long after receiving the packet. In a letter of thanks, Kevorkian wrote,

> "As emphasized in my papers, I neither support nor oppose capital punishment; my only contention is that wherever and whenever it is to be irrevocably used, some condemned persons should be given this choice. Even opponents to the death penalty (such as Dr. [Hans] Selye and yourself) have admitted that this proposal is a fair compromise and desirable if the death penalty is not to be totally abolished."

Later in the letter, Kevorkian requested written backing from Pauling, for use in an upcoming hearing to be held by the Florida legislature. Pauling responded in kind, writing "I support your proposal about an alternative method of capital punishment."

Three and a half years later, in November 1976, Kevorkian wrote again. At that point, capital punishment in the U.S. had been suspended but, as Kevorkian noted, "it seems that we are on the verge of seeing a resumption of executions, perhaps soon in Utah. I plan to resume my campaign and would like to know if you still endorse the idea and still allow me to cite your endorsement publicly."

By then, Pauling was starting to have his doubts. In a letter written promptly upon his receipt of Kevorkian's request, he wrote, "I am so busy at the present time that it would be difficult for me to look into the matter, and in fact my ideas have changed somewhat. I think that the best form of execution, if people are to be executed, is the one that causes the person the least suffering. I am not sure that this idea is compatible with your proposal." No further record of exchanges between Kevorkian and Pauling remains extant in Pauling's papers.

The Linus Pauling Institute of Science and Medicine

Chapter **20**

By Adam LaMascus

Group photo of the LPISM staff, January 1989.

In 1969 Linus Pauling was given his own laboratory at Stanford University for his work on schizophrenia, where Art Robinson, a colleague and fellow researcher, joined him. By 1972 Pauling and Robinson had decided that their Stanford facility was no longer sufficient for their growing research program, and in May of that year Pauling requested that the university construct a new building to house an expanded lab for him to use.

The institution hesitated in responding, as its administration was somewhat wary of Pauling at that time, given the controversy that surrounded both his scientific interests and political activism. Pauling had recently co-authored divisive work with the Scottish physician Ewan Cameron about vitamin C and its usefulness in treating cancer, research that alienated him from much of the medical community. For its part, Stanford was very unsure about the wisdom of giving him a new lab to further that work.

Likewise, Pauling was pushing hard and visibly for global peace in the middle of the Cold War, activities which had long caused many people to suspect that he harbored Communist sympathies. Almost as if to verify that accusation in their minds, he had been awarded the Lenin Peace Prize by the Soviet Union in 1970. Pauling was a harsh and public critic of the war in Vietnam, President Richard Nixon, US nuclear policy, and US foreign policy, which only served to legitimize some people's doubts about Pauling's loyalty to the United States. And if that wasn't enough, Pauling had also publicly protested Stanford's firing of a tenured professor known to have leftist, anti-war leanings. Stanford took all of these details into consideration and decided to deny his request.

In response, Robinson suggested that Pauling step down from the Stanford faculty and move their research off campus, which they did, relocating in early 1973 to a building that Robinson had found nearby. Their new home was at 2700 Sand Hill Road in Menlo Park, across the street from the Stanford Linear Accelerator. The space was in an office building shared with Kemper Insurance, a location never designed to accommodate scientific research. Nonetheless, Pauling and Robinson adapted, figuring out how to fit their operation into a footprint of less than 20,000 square feet, and in an area designed to hold desks and chairs instead of wet labs.

Freed from their affiliation with Stanford University, Pauling, Robinson and Keene Dimick — a biochemist who agreed to help pay the rent for their new quarters — decided to form a brand new institute.

On May 15, 1973, the Institute of Orthomolecular Medicine was founded as a California non-profit research corporation with the stated goal of researching biology and medicine.

(In late 1973, Pauling decided to sever all of his professional ties with Stanford, equally annoyed with the university as it seemed to be of him. However, he still retained a number of good friends amongst the faculty there, with whom he maintained close ties and corresponded frequently.)

Immediately the Institute began attacking a problem that would plague it for most of its existence: funding. Pauling and Robinson started by lobbying many of their friends and associates for money, trying to entice them in part by offering largely honorific positions on the Board of Associates of their new organization. Over 30 people agreed to help, including Francis Crick, Maurice Wilkins and a number of other Nobel Laureates. The funds generated weren't princely, but they were enough to get operations started.

On the research end, the Institute began by continuing the vitamin C and metabolic profiling program that had occupied Pauling and Robinson at Stanford. The new labs contained a large number of animal experiments, which required the bulk of the very limited space. Wedged in as well was a Volcano Source Field Ionization Mass Spectrometer, a new device being developed by Bill Aberth, a recent hire who had previously worked for SRI International.

Pauling wanted to help people with his research on vitamin C, and in 1974 he opened a small outpatient clinic run by John Francis "Frank" Catchpool, a doctor whom Pauling had met in 1959 while visiting Albert Schweitzer's medical hospital in Africa. Unlike Schweitzer's venture, the Institute's clinic was immediately beset with major problems; namely, liability was too high and funding was too low. Additionally, because vitamin C was cheaper than other medication, the clinic routinely found itself overwhelmed by destitute patients. The staff often ended up working for free, as people would arrive who were too poor to pay for anything, but Cameron and others continued to help out of sympathy and a desire to do good. Due to severe limitations on resources, the clinic closed in 1975, just eight months after it opened.

As 1974 progressed, the Institute's funding problems became increasingly ominous, exacerbated by the fact that Robinson was one of the only people on staff who was adept at fundraising. In July the Institute's Board of Trustees decided to start addressing the problem by renaming the Institute of Orthomolecular Medicine. They came to the conclusion that the term "orthomolecular" was not only too difficult to explain to most potential donors, but that the word itself had been tainted by a recent barrage of attacks on vitamin C by the American Psychological Association. They also decided that having Pauling's name attached to the Institute would help with funding, due to his international fame and respect.

On July 26, the Institute was officially renamed the Linus Pauling Institute of Science and Medicine (LPISM). The board's choice proved to be a good one, and more funding began to come in. Even still, it wasn't enough, and for much of 1974 LPISM was kept afloat through financing from Pauling and Robinson's personal accounts. Ultimately this too proved to be insufficient, and by early 1975 LPISM was in danger of succumbing to bankruptcy. Desperate, the Institute's administrators were forced to issue pay cuts for all employees, but did so on a sliding scale designed to minimize the impact on workers with already low salaries.

It was at this point that Robinson confronted Pauling, accusing him of neglecting his work as president. Pauling had in fact been derelict in certain respects, boasting to the press at one point: "I don't waste time on needless details." Unfortunately, many of these details weren't needless, and his administrative inattention was harming the Institute. Pauling asked Robinson if he thought he could do better, to which Robinson replied that he could. Pauling responded by making Robinson President and Director of LPISM. Still the financial issues became worse: Pauling donated 75% of his income to LPISM for the first half of 1975, and then 100% of his income in the second half of the year.

In the meantime, the Institute's staff began experimenting on hairless mice. For a week before a test, they would feed the mice food full of vitamins C and E. Next they began irradiating them with ultraviolet light to produce skin cancer, and then observed the effects of the high-vitamin diet on cancer growth. As a result of this research, Pauling developed a cocktail of vitamins, which he put into a single dose that he called the "Linus Pauling Super Pill." The Institute considered marketing the pill to raise additional funds, but that plan fell through and the Super Pill was relegated to a formula in a filing cabinet.

*

By the end of 1975, LPISM found itself teetering on the brink of financial collapse and everyone involved realized that the organization could not hope to succeed if business was to continue as usual. An important change came in March 1976, when Rick Hicks was hired to help with fundraising. This move quickly proved auspicious as Hicks interjected new energy and ideas into the search for additional income. One of the Institute's most important employees for many years, Hicks quickly deployed his talents to help mitigate the extreme financial problems of 1975. With Hicks on board, the Institute was only treading water, but the imminent threat of drowning had lessened.

Later in 1976, Pauling published *Vitamin C, the Common Cold, and the Flu*, an updated version of his successful earlier book, *Vitamin C and the Common Cold*. The book sold reasonably well and managed to bring some needed cash into LPISM, though not as much as the organization had hoped. Around this time, Cameron and Pauling published another paper on vitamin C and cancer, which stated that terminal cancer patients lived longer and enjoyed a higher quality of life when given supplemental vitamin C.

Just as conditions were starting to improve, the Institute suffered another major setback when investigators from the National Cancer Institute (NCI) visited LPISM to check on the progress of its hairless mice skin cancer study. In their report, the investigators frowned upon the Institute's tiny staff, what it deemed to be bad management by its administrators, and Pauling's frequent absences. Their harsh judgement badly damaged the Institute's grant funding prospects for many years afterward, and forced LPISM to redouble its efforts to secure new sources of private funding.

On the research front, Art Robinson began working with Kaiser-Permanente to set up a national sample bank stocked with tens of thousands of blood and urine samples that LPISM could utilize for better research. The collaboration was moving along smoothly until Kaiser realized that the bill for the project would be in the neighborhood of $5.8 million, at which point they decided to scrap the deal. In 1977, a year after negotiations had started, Robinson was informed that the plan was rejected because it was too ambitious and vague.

Robinson wasn't the only one receiving that response. Pauling had been seeking grant funding from the NCI from the moment that its investigators' report had been released. He reasoned that if the NCI was complaining that LPISM was too small, then they should give him more money to expand. By 1977 he had been turned down four times by the organization, on each occasion under the auspices of his proposals also being too ambitious and vague.

In light of the bad press that accompanied the NCI's dismissals, Pauling sent copies of his proposals and rejection letters to 24 members of Congress, including the senators heading the committees on health and nutrition, Ted Kennedy and George McGovern respectively. He received no response, and in turn asked his lawyer if he could sue the NCI for bias. His lawyer advised him that there was no legal precedent for such a case, and that the chances of successfully suing a federal agency were "less than slim."

In early 1977, LPISM hired two people, both of whom ended up making major contributions to the Institute. The first was Emile Zuckerkandl, recruited to lead their research on genetics. Zuckerkandl brought a different type of personality to the Institute. He and his family had left their home in Austria shortly before the onset of World War II, as they were wealthy and Jewish. In fleeing they had managed to bring with them some of their rare art collection, which Zuckerkandl would show to his coworkers at LPISM from time to time. A renowned scientist,

Zuckerkandl had worked previously with Pauling at Caltech, collaborating and co-developing a theory of molecular evolution.

The second hire was Steve Lawson, who was brought on board to assist with direct-mail fundraising solicitations. Lawson's position was very much at the entry level. When he was hired he was not associated with the Institute in any way, and his only concrete knowledge of Linus Pauling harkened back to a chapter during his student days at Stanford University, when he saw Pauling protesting the aforementioned firing of a tenured instructor.

Around this time, the NCI also brought someone in: a new head named Vincent DeVita. He was more open-minded about Pauling and his vitamin C work, and was aware of how much public support Pauling had accumulated. As a result he consented to speak with Pauling and, after a multi-hour meeting, agreed to set up a conclusive clinical trial on vitamin C and cancer. When asked about the apparent reversal of the NCI's opinion of Pauling's work, DeVita responded that Pauling "can be a very persuasive man." By March, DeVita had arranged for the trials to be carried out at the prestigious Mayo Clinic, headed by Dr. Charles G. Moertel. Pauling and DeVita met again in April to discuss the details of how the trial would be conducted.

With the Moertel study forthcoming, circumstances appeared to be improving across the board for the Institute. Buoyed by an apparent increase in support for his research, Pauling expanded his program and, in 1977, launched a new series of tests on 600 hairless mice to determine the effect of vitamins on tumor growth. Additionally, LPISM managed to accumulate enough capital to hire a professional direct-mail company for fundraising, and was able to place a number of successful advertisements in major financial periodicals including *Barron's* and the *Wall Street Journal*. With these changes, LPISM saw a hefty boost in its incoming funds.

The Institute's initial successes with direct mail prompted it to invest more heavily in this strategy. The move proved to be very effective with non-governmental donations increasing from 50% of LPISM's funding to 85%, including almost $1.5 million in private donations in 1978 alone. For the first time in its short history, LPISM wasn't suffering from financial hardships.

*

This wave of good fortune, however, carried with it unforeseen negative consequences. In particular, Rick Hicks and Art Robinson began to come into conflict over the best way to invest the Institute's sudden surplus. Robinson suggested that LPISM move to Oregon — which had recently announced "Linus Pauling Day" in honor of its native son — and build a campus of its own. The idea was not popular with many staff, most of whom did not want to leave the Bay Area.

At the same time, Robinson began cultivating ties with the Orthomolecular Research Institute (ORI) in Santa Cruz, California, which was headed by Arnold Hunsberger. Pauling was not pleased with this, as he felt Hunsberger's research hypotheses to be off the mark. Pauling had also met Hunsberger and said that his impression was "not a very favorable one."

As Robinson continued to press for closer ties between LPISM and ORI, tensions grew between him and Pauling. Pauling was further angered when he learned that Robinson had begun to tailor experiments in accordance with Hunsberger's ideas without first consulting Pauling. When confronted, Robinson defended his decision and redoubled his arguments for collaboration. The relationship soured from there and morale at LPISM plummeted as disagreements between Pauling and Robinson became more common.

A breaking point arrived in June 1978, when Pauling issued a memorandum to Robinson, ordering him to consult the Executive Committee — comprised of Pauling, Robinson, and Hicks — before making "any important decisions." Robinson responded by immediately firing Hicks. Pauling, in turn, overruled the termination and demanded Robinson's resignation within 30 days. He then circulated an additional memorandum informing Institute staff that he had stripped Robinson of his position, and that the staff was to disregard all further instructions from

Robinson. The next day, employees arrived at work to find a responding memorandum from Robinson declaring that he was still the president, that neither Pauling nor Hicks had the authority to relieve him of his duties, and that he would not resign.

The Board of Trustees met in mid-July to try and settle the dispute. They decided to place Robinson on a 30-day leave of absence, empowered Pauling with all executive authority, and told him to resolve the issue. On August 15, with Robinson's leave expired, Pauling was elected President and Director of LPISM. On August 16, Pauling promptly informed Robinson that he was taking over all of Robinson's research, Emile Zuckerkandl was being appointed Vice-Director, and that Robinson was fired.

With Robinson ushered out, LPISM attempted to consolidate and return to normal. Pauling asked Steve Lawson to assume a portion of Robinson's research portfolio, a request to which Lawson consented. Over the course of the year, Lawson had become less connected with the financial arm of LPISM and more involved with its scientific work. Zuckerkandl also tasked Lawson with setting up a cell culture facility where the two would conduct research on the differences between primary and metastatic cancer cells, as revealed by protein profiling. In doing so, Lawson worked closely with colleagues at UC San Diego, University of Colorado, and SRI International. He was later joined on the project by Stewart McGuire, Eddy Metz, and Mark Peck, all fellow employees at LPISM.

Robinson, meanwhile, did not take his firing lightly and on August 25, LPISM was informed that he was suing the organization for $25.5 million, alleging breach of contract and unlawful termination, among other charges. LPISM's lawyers began gearing up for a serious legal battle, standing firm in their conviction that the Institute had done nothing wrong.

In the midst of it all, the Institute's vitamin C research continued. In early October 1978, Pauling convinced Ewan Cameron to accept a one-year appointment to LPISM while the two worked on a book about vitamin C and cancer. Additionally, Pauling, Cameron, Lawson, and their coworker Alan Sheets began an experiment to determine the effects of vitamin C on chemotherapeutic drugs. The research took the form of a toxicology experiment in which multiple groups of fish were subjected to chemotherapeutic agents in their water, after which various groups were given different amounts of vitamin C while the research team observed the results.

The year 1979 started with good news. LPISM was informed by Hoffmann-LaRoche, the world's largest producer of vitamin C, that they had seen sales more than double during the 1970s, and that they fully recognized that Pauling was the cause. As a thank you, they had decided to donate $100,000 a year to the Institute.

But the happy days were not to last long. In April, LPISM received an advanced release of the results of the Mayo Clinic's major study on the treatment of cancer with ascorbic acid. Its primary investigator, Charles Moertel, had concluded that vitamin C did absolutely nothing to help cancer patients. Pauling was stunned and immediately began writing to Moertel to discuss the study in detail.

Then, over the summer, Art Robinson filed six more charges against LPISM and Pauling, bringing the total number of suits to eight and the total requested damages to $67.4 million. The year-long and highly publicized legal battle was greatly hurting LPISM's reputation, and the Institute noticed a subsequent decrease in the donor funds flowing their way.

Things went from bad to worse when, on September 27, the *New York Times* published the Mayo Clinic study, leading with its conclusion that vitamin C was useless in treating cancer. Pauling immediately responded by pointing out that the patients involved in the test were undergoing cytotoxic chemotherapy, which he felt crippled their immune systems. He also asserted that the trial was not conducted for a lengthy enough period to develop accurate results.

Moertel returned fire, defending his results and implying that Pauling was fanatical in his zeal for vitamin C and his refusal to acknowledge the truth. From there, Pauling and Moertel began exchanging volleys in public, writing articles and giving interviews that attacked the research and competence of the other. Unfortunately for Pauling, he took the worst of it, as many people began to agree with Moertel, thinking Pauling to be too enamored with vitamin C to see any negatives. Funding plummeted as donations shrank and LPISM began finding large numbers of grants rejected outright with no chance for an appeal.

But Pauling refused to give up. Shortly after the *New York Times* article was released, he and Cameron published their book, *Cancer and Vitamin C*. Pauling personally bought 16,000 copies of the publication and mailed them to every member of Congress and to myriad other physicians and researchers. This action helped Pauling's cause significantly as many of the recipients read the book, or at least glanced through it, and even those recipients who didn't read the text were made more aware of Pauling and his research. Likewise, in the marketplace the book sold well despite the bad reception it had received from professional reviewers; the public seemed interested in Pauling and Cameron's ideas.

In light of this, NCI head DeVita agreed to a second round of trials. However, in doing so DeVita once again chose the Mayo Clinic to host the trials and selected Moertel to lead them. Pauling was furious with these decisions, an understandable point of view considering that he and Moertel had spent the past few months publicly accusing each other of being incompetent. Pauling was also now without his co-author: their book completed, Ewan Cameron had returned to Scotland to fulfill his duties at Vale of Leven Hospital. Before leaving, he was appointed as Research Professor at LPISM for a period of five years.

*

In the spring of 1980, amidst a swirl of funding difficulties and legal actions, Emile Zuckerkandl was named President and Director of LPISM. He quickly began working to expand LPISM into a more wide-ranging organization with a growing interest in cellular research. His leadership style was quite different from the Institute's previous presidents, but the staff liked him and generally supported his ideas.

By this point, born of need, some of Pauling's decisions related to the Institute began to take on a more dubious quality. Notably, Pauling was contacted by, and began meeting with, a man named Ryoichi Sasakawa to discuss future collaboration plans and possible donations. Sasakawa was a world-renowned philanthropist and famous businessman who had single-handedly introduced and popularized motorboat racing in Japan. Sasakawa was also very controversial. An avowed fascist, he was an admirer of Benito Mussolini and a political strongman who had been charged with war crimes for his activities in support of the Japanese government during World War II. He was also very wealthy and Pauling's connection to Sasakawa would grow over time.

The summer and early fall of 1980 were largely preoccupied with the Art Robinson suits and fundraising. In August, LPISM finally received some good news: the National Science Foundation had awarded the Institute a grant of $40,000 a year for two years to support research on the structure of molecules and complex ions containing transition metals. This provided a much needed boost, as finances were suffering greatly from the Mayo trials fallout and the ongoing legal wrangling with Robinson.

The year ended somewhat stressfully when, in December, LPISM was forced to move from Menlo Park to 440 Page Mill Road in Palo Alto. The landlord of their building in Menlo Park had evicted all of his tenants while he was making structural repairs to the facility. Once completed, he decided not to welcome LPISM back, instead inviting more profitable companies to take their spot. The new building in Palo Alto was dramatically bigger and less expensive; it was also quite a bit shabbier, in part because it was made out of cinderblocks. Employee Alan Sheets was able to help save the Institute a lot of money during the transition, as his father was a professional mover. As such, the Sheets family helped LPISM move itself instead of hiring the process out to a company.

The dawn of 1981 brought with it major financial relief for LPISM. After eight failed tries over eight long years, NCI finally agreed to fund a component of LPISM's program — a two-year grant of $204,000 to research the effects of vitamin C on breast cancer in mice. At about the same time, Sasakawa's company, the Japanese Shipbuilding Foundation, pledged $5 million to the Institute over the following ten years. As part of the deal, LPISM began working with the organization to create the Sasakawa Aging Research Center, which was set up as a satellite facility on Porter Drive. Later in the 1980s, the building at Porter Drive suffered a major roof leak which destroyed thousands of pages of research and documentation. Biographer Thomas Hager has noted that LPISM successfully sued the landlord for neglecting to maintain the building.

Despite this influx of new cash, the close of 1981 proved to be an awful time for Linus Pauling and LPISM. In August, Ava Helen Pauling's recurrent stomach cancer was declared inoperable and on December 7, after struggling with the disease for five years and three months, Ava Helen died. Her husband was absolutely devastated, and the LPISM staff was greatly saddened by the loss as well. Pauling understandably experienced great difficulty in coping with the death of his wife of nearly 60 years, and for a period he effectively ceased to be involved in LPISM except in the most cursory of ways, choosing instead to spend much of his time alone at his ranch in Big Sur, California.

The year that followed was, unsurprisingly, a tough one. Pauling remained in mourning, the Robinson suits dragged on, and the Institute's fundraisers still struggled to overcome the fallout from the Mayo trials. The NCI grant and Sasakawa donation helped to keep operations running, as did some of the revenue from Pauling and Cameron's book. In the summer of 1982, Pauling took a trip throughout the Pacific Northwest where he visited many of his and Ava Helen's favorite spots, as well as the cemetery where his maternal grandfather Linus Wilson Darling rested. The trip helped bring closure and by the fall he became active at the Institute again.

In February 1983, the lawsuits with Arthur Robinson finally ended, with LPISM paying an out-of-court settlement of $575,000. The Institute adamantly maintained no wrongdoing, instead acknowledging the fiscal prudence of settling as opposed to prolonging the court battle, which was nearly five years old by that point.

Their legal problems resolved, LPISM fundraisers redoubled their efforts to regain momentum, as the lawsuits had drained them of resources. Past fundraising techniques were unable to generate much steam, so Rick Hicks began cultivating relationships with individual, extremely wealthy donors, notably Armand Hammer, Danny Kaye and the aforementioned Ryoichi Sasakawa. As a part of this strategy, LPISM began annually awarding individuals — typically major donors — the Linus Pauling Medal for Humanitarianism. Sasakawa was its first recipient.

In November 1983, LPISM researchers announced that they had discovered a new type of chemical bond that mimicked the bond believed to exist between bulk metals. This was a fairly important discovery, and also helped restore some measure of favorable public opinion as people saw the good work that LPISM was doing. The announcement also reminded potential donors that LPISM wasn't just about vitamin C research. The next year, Pauling received the extremely prestigious Joseph Priestley Medal from the American Chemical Society for his lifetime of work and dedication to the field of chemistry.

But the controversy over vitamin C was never far from the Institute and more arrived in a hurry when, on January 2, 1985, the Mayo Clinic released the results of its second set of trials. The Institute was given no warning of this release nor opportunity to read the results in advance. This infuriated Pauling who saw it as an obvious insult levied by Moertel, the study's principal investigator.

Perhaps unsurprisingly, Moertel announced that the study had reaffirmed his earlier assertion that vitamin C was useless in treating cancer. Upon reading the report though, Pauling deduced that Moertel hadn't actually examined Ewan Cameron's papers, the very studies he was supposed to be replicating. Among other deviations, the amount of vitamin C used in the Mayo trials was lower than in Cameron's studies, the amount of time that patients had been given vitamin C was shorter, and patients were given vitamin C orally instead of intravenously. Both Pauling and Cameron publicly branded the Mayo report as "fraudulent" and angrily decried the false assertion that Moertel had closely replicated their work.

Many journals and newspapers refused to publish Pauling and Cameron's rebuttals, or published them months after they were submitted, such that the responses were no longer relevant. As a result, LPISM suffered still more financial hardship as public opinion once again swung away from the Institute and many people stopped donating. The direct-mail appeals that had been so successful in years past were only bringing in 25% of what they had garnered a few months previously.

By 1986 LPISM was struggling with funding and also public awareness; the second Mayo Clinic trial seemed to have largely sealed public opinion on vitamin C research. But Pauling was still convinced that vitamin C had more merit than was being considered, and in support of this cause he published *How to Live Longer and Feel Better*, written for a lay audience. The book was applauded by critics and sold well.

For the Institute, the book's successes were manifold, as it provided a morale boost to LPISM staff, brought in sorely needed funds, and dramatically raised awareness of the organization and its activities. In the wake of its publication, Cameron and fellow LPISM employee Fred Stitt found themselves swamped with phone calls and letters about health questions and recommendations. They quickly developed a standardized health information packet that they would mail out to people who had made more generic inquiries.

Nonetheless, as always, controversy was hovering over the Institute like a thunderhead. In 1987 Institute staffer Raxit Jariwalla began to research the effect of vitamin C on HIV/AIDS treatment. After a short period of time, Pauling became interested in the research and eventually Cameron did as well. Pauling began advocating increased usage of vitamin C in treating what seemed to be an incurable disease, and the response was immediate and dramatic. Local donations increased, as the Bay Area was particularly sensitive to the hazards posed by HIV/AIDS.

However, at the very same time, other sources of funding dropped as numerous groups and individuals pulled their support, stating that HIV/AIDS was a "moral disease."

Through it all, the Institute continued to follow Zuckerkandl's lead in expanding its research into areas outside the realm of orthomolecular medicine. In 1987 researchers began extensive work on protein profiling and the effect of phytic acid in cancer prevention, a program that was more or less entirely supported by a philanthropist based in New York.

The Institute also began investigating superconductivity in 1988. In particular, Pauling hoped to develop a room-temperature supercon-ductor which he could then market as a stable revenue source for the Institute. A group including Zuckerkandl, Lawson, Pauling and even Cameron began working on the project, which utilized a material made out of borosilicate glass and tin. The process involved using a blowtorch and an inverted bicycle with its tires taken off the wheels. Pauling would often come down to the labs and help with the experimentation — it was, in fact, the last bench research project he actively participated in. The process worked and the material was developed according to Paul-ing's specifications. He received a patent for it early in 1989, and imme-diately began trying to market it, though ultimately without success.

*

As the 1990s approached, LPISM soldiered on as best as it could. One who would help define the decade was Matthias Rath, a young, charis-matic and intelligent German physician who had a passion for vitamin C and cardiovascular health. He had met Pauling on numerous occa-sions, and in 1989 Pauling invited him to join the LPISM staff. Rath was charming and popular with many of his colleagues. However, Paul-ing's oldest son, Linus Jr. — a long-time Institute board member — took caution, noting in a 2012 interview his concern that Pauling would offer a position of importance to somebody whom he felt to be very inexperienced.

Two other major events occurred in 1990: Pauling and colleague Zelek Herman developed a new method to analyze clinical trial data, and NCI installed a new president, Samuel Broder. Pauling immediately began corresponding with Broder, and eventually convinced him to reopen the case for vitamin C as both a preventative and a treatment for cancer. This resulted in an international conference held in Washington, D.C. in 1991 and sponsored by the NCI. It was titled "Ascorbic Acid: Biological Functions in Relations to Cancer." Pauling was the obvious candidate for keynote speaker and he later said of the conference, "It was great! A great affair! Very exciting!"

At this same time, Pauling created a new position at LPISM for Rath, who was named the first Director of Cardiovascular Research. With this, Linus Jr. became even more concerned. Increasingly, he came to question his father's administrative acumen and began taking steps to assume a more active role in the management of the Institute, despite the fact that he lived in Hawaii.

Another big change was on the horizon as well. The city of Palo Alto was planning to change their zoning laws in an effort to increase residency, and informed LPISM that they had three years to find a new home. The Institute believed that the time allotted them was insufficient, and they began a campaign to delay the eviction. Staff set up card tables in front of businesses, dispersing flyers and circulating a petition to keep LPISM at the Page Mill Road location.

The positive response that they received from the locals was staggering and gave the Institute some measure of leverage in their conversations with the city. At one point, Steve Lawson was called before the city council, and during the hearing one member said that she didn't want to read in the *New York Times* that Palo Alto had kicked LPISM out of town. Eventually the council informed LPISM that the zoning law changes were still going to go through, but that the Institute would be granted more time to plan and relocate.

On the research front, after almost two years of marketing Pauling's superconductor domestically with no leads, Rick Hicks decided to look abroad for a buyer (see Chapter 18). He contacted parties all over Europe and Asia, and one day a man showed up at the office to inquire about superconductor sales. He identified himself as an employee of the Central Intelligence Agency, which had taken an interest as to why LPISM was trying to sell this research internationally, especially in Japan, instead of on the U.S. market.

Hicks was away from the office at the time, but other employees were able to explain how he had tried unsuccessfully to sell it domestically first. Lawson later recalled the experience as having been a jarring one. Unfortunately for LPISM, they also failed to sell the superconductor abroad and, due to an oversight, misplaced the paperwork required to pay the royalty fee needed to maintain the patent, which they lost as a result.

While this was going on, Pauling and Rath published a paper defining vitamin C deficiency as the major cause of cardiovascular disease. It immediately caused controversy, but the authors stood behind their work and continued to pursue it. Once again, concerns about Pauling's infatuation with ascorbic acid flared in the scientific community.

Another blow to the Institute's fortunes was delivered on March 21, 1991, when Ewan Cameron died. His passing rocked the staff and morale plummeted. Shortly afterward, Pauling was diagnosed with prostate cancer and had to undergo surgery. On top of all of this, the fiscal report for the end of 1991 showed that LPISM was hundreds of thousands of dollars in debt. Workers remained loyal however, and numerous employees volunteered to suspend retirement contributions or work at reduced pay to keep the Institute afloat. Despite this, LPISM was still forced to cut their staff in half by early 1992.

Meanwhile, Pauling and Rath continued to promote vitamin C for cardiovascular disease prevention and treatment, and despite continuing doubts about their individual claims, they began to see more support as the medical community gradually realized that it had been underestimating the value of vitamin C for decades. As their work progressed, Rath's connection to Pauling strengthened.

In the spring of 1992, more change was clearly afoot when Emile Zuckerkandl's contract with LPISM was not renewed. This was a controversial move, as Zuckerkandl was well-liked and respected by the staff. After his departure, he founded his own institute, the Institute of Medical Molecular Sciences (IMMS). He asked the Board of LPISM if he could lease space within LPISM for his new IMMS, a request that was granted.

Additionally, Zuckerkandl invited many of the LPISM staff who had been laid off to join IMMS. When he received news that Zuckerkandl was leaving, Rick Hicks, who by now was the Vice President for Financial Affairs, submitted his resignation as well. He had worked very closely with Zuckerkandl and wanted to follow him to other business ventures. The board was surprised by Hicks's resignation and the Institute didn't want to lose its affiliation with him completely, so they elected him to their body to keep him tangentially involved in LPISM. Happily, one of Hicks's last acts as an employee was to inform the Board that the estate of Carl L. Swadener had been bequeathed to the Institute, with a valuation of $2–3 million.

Linus Pauling Jr. was elected as the next Institute President, replacing Zuckerkandl, and the organization that he took over was in grim shape, despite the windfall from the Swadener estate. As he assumed his new office, one of his top priorities was Matthias Rath. Amidst the recent shuffle, Linus Pauling had appointed Rath as Hicks's replacement, and at the same time the two had founded the Linus Pauling Heart Foundation, a separate and parallel organization to LPISM designed to focus on the Pauling-Rath cardiovascular disease research. These decisions were a source of concern to the board and much of the staff, who were unsure if the Heart Foundation would be a competitor to the Institute, an arm of the Institute, or a supporting organization to the Institute.

Overwhelmed by work, facing a serious illness and feeling his age, Linus Pauling officially retired from his leadership role at LPISM on July 23, 1992. In the wake of this announcement, the board elected Steve Lawson as Executive Officer of the Institute, named Pauling its Research Director, and Linus Pauling Jr. the Chairman of the Board. Linus Jr. immediately assumed a strong leadership role and, working closely with Lawson, aggressively pursued actions to solve the Institute's numerous problems.

The two quickly decided that attaching LPISM to a university offered the best chance for its survival. At the same time, they realized that LPISM had become bloated and that they needed to pare back the organization's non-orthomolecular research, which had largely been initiated and expanded under Zuckerkandl's watch. While Linus Jr. and Lawson both agreed that the research was worthwhile, they also realized that the Institute simply lacked the capacity to maintain it. Zuckerkandl had remained close to LPISM, and when almost all of his research programs were cut, he asked the researchers overseeing these programs to resign from LPISM and join IMMS, which many did.

While this was happening, tensions were mounting between Pauling, Linus Jr. and Matthias Rath. Pauling was informed that Rath had created an office for the Heart Foundation that was separate from LPISM, and that he had done so without permission or proper communication. Pauling criticized Rath aloud for this decision, which only inflamed the situation. From there, the Pauling-Rath relationship soured in dramatic fashion. In July, Rath was spending great amounts of time at Pauling's home, and they frequently exchanged letters expressing a close friendship. By August they were hardly on speaking terms, and Rath was ultimately expelled from the Institute, asked to resign over a dispute involving intellectual property rights.

*

In 1993 Steve Lawson's title was changed from Executive Officer to Chief Executive Officer, though his duties effectively remained the same. At the request of Pauling, one of Lawson's first actions was to legally dissolve the Linus Pauling Heart Foundation. In doing so, he dismissed all of the Heart Foundation's employees and transferred the entity's assets to LPISM. At the same time, the Palo Alto zoning law changes of which the Institute had been warned went through; the Institute finally needed to devise a solid idea of where it was going to move.

In the meantime, Lawson, looking to alleviate LPISM's perpetual financial problems, began negotiating with the Elizabeth Arden beauty company on a deal that he hoped would greatly enhance the Institute's well-being. Arden and its parent company, Unilever, were seeking research support and eventual endorsement from LPISM for an upcoming line of skin care products, which were infused with vitamin C. Lawson was interested in both the financial and advertising benefits that might come from this deal, as the Institute badly needed to increase its exposure to a younger and wealthier audience.

The conversation was proceeding smoothly until Arden installed a new president, a time period during which the release of the new products was managed in an unexpected way. This person only remained president for a short period, but the damage had been done. As a result, the deal between LPISM and Elizabeth Arden proved dramatically less prosperous than Lawson had hoped.

With the Arden deal scrapped, the Institute's administration encountered more bad news when they received notice that Matthias Rath was suing LPISM, alleging interference with his business practices. Following his departure from the Institute, Rath had encountered difficulty summoning financial support for his vitamin C work, with some people assuming that he was trying to claim credit for Pauling's research. One magazine in particular had published an extensive article on Pauling's interest in vitamin C and cardiovascular disease and hadn't even mentioned Rath. LPISM asserted that Pauling had acknowledged Rath's contributions in his interviews and that the Institute had no control over what various media outlets published. The lawsuit proceeded nonetheless.

The year 1994 got off to a very bad start. It was around this time that Pauling's health began to deteriorate markedly and he was forced to undergo treatment for his resurgent cancer, which had spread to his liver. At the same time, the lawsuit with Rath intensified while Pauling spent more and more time away from the office, choosing instead the tranquility of his ranch at Big Sur. By the summer, Rath's lawyers were visiting Pauling's bedside to try and hash out an agreement. For the Institute, most of the year was spent dealing with these two major issues.

Finally, on August 19, 1994, Pauling died at his ranch. The institute that he helped create and that bore his name instantly felt an intense drain on its spirits. Lawson recalled employees sobbing in their workspaces and noted that many staffers felt directionless, unsure what would become of LPISM without its namesake. Ironically, the organization's financial problems were a bit relieved by this turn of events, as a flood of memorial donations soon came in.

From this moment of darkness, the situation pretty quickly started to improve. Amidst the tumult, the Institute went through with a scheduled September conference, titled "The Therapeutic Potential of Biological Antioxidants." Many people attended — more than were expected — and the audience was thrilled with the content presented, responding very enthusiastically. In turn, more donations and support began to flow into the Institute's coffers.

At the same time, the Institute received notification that another estate of consequence — the Finney estate — had been left to LPISM. This new revenue source, combined with the Swadener gift, allowed LPISM to effectively pay off its debts and even establish a small endowment to support the Institute's relocation. Coincidentally and almost simultaneously, a large number of bequeaths and other gifts began pouring in, largely from donors who had been cultivated years before by Rick Hicks.

The financial situation suddenly and vastly improved, Lawson and Linus Jr. started hunting for a new location for LPISM. They began by contacting universities all over the U.S., with decidedly mixed results. Frustrated, the Institute's board briefly considered closing down LPISM in favor of establishing a memorial chair at Caltech or Stanford.

However, Oregon State University (OSU) eventually came forward and requested that LPISM relocate to Corvallis. In stating its case, the university stressed its historical connection with Pauling, as OSU was his alma mater and home to his papers, which were housed in the university library's Special Collections. OSU's argument also pointed out that its existing chemistry, health, and biomedical programs nicely complemented LPISM and its research. Linus Jr. and Lawson agreed, and decided to move the Institute to the heart of the mid-Willamette Valley.

*

Even though the reorganization of the Institute after Emile Zuckerkandl's departure had shrunk its staff from 75 to 50, it was still determined that LPISM was too big to move to Oregon in its then-current size. For one, many of its development obligations would no longer need to be assumed by Institute staffers, as the OSU Foundation had agreed to lead fundraising efforts. Other staffing redundancies were also quickly becoming apparent.

As the Institute's CEO, Steve Lawson began to meet regularly with OSU's Dean of Research, Dick Scanlan, with the two carefully studying staff lists, deciding who and what was most likely to succeed at OSU. Eventually it was agreed that LPISM would move to OSU with a skeleton holdover staff of five people: Lawson, Conor MacEvilly (biochemist), Vadim Ivanov (cardiovascular disease researcher), Svetlana Ivanova (Ivanov's wife and research partner), and Waheed Roomi (researcher focusing on the cytotoxic molaity of vitamin C derivatives) would come to Oregon.

In preparing for the move, Lawson worked closely with Scanlan and OSU President John Byrne to hammer out the specifics of how to integrate LPISM into OSU. In 1995 Linus Pauling Jr., Lawson, and incoming OSU President Paul Risser all signed a Memorandum of Understanding that laid out how everything would be transferred to OSU, and how LPISM would be legally dissolved as a separate entity. OSU promised to provide the Institute with administrative and laboratory space on the fifth floor of a science building, Weniger Hall, which had just been renovated. The university also pledged additional funding for salary lines, and to work toward eventually housing LPISM in its own building should it someday outgrow Weniger Hall.

The big move was made in July 1996. LPISM was able to bring with it an endowment of $1.5 million, which the state of Oregon agreed to match. As they prepared for relocation, the remaining staffers purged much of their material: Lawson estimated that they filled two full-sized dumpsters per week as the move progressed.

Upon arrival the Linus Pauling Institute (LPI) was created as a separate legal entity from LPISM, which continued to exist as a shell company for several years afterward. LPISM needed to continue to live as many bequests had been specifically made out to LPISM. There was also the issue of standing lawsuits from Matthias Rath and another former staffer who was suing LPISM for wrongful termination. Due to these legal imperatives, and despite the fact that, by 1996, it had ceased to exist on anything but paper, LPISM was not finally dissolved until the mid-2000s.

Once settled in Corvallis, the Institute's fortunes continued to improve. Importantly, the financial problems that had plagued the Institute for all of its life basically vanished. Regular influxes of donations coupled with residence at OSU saved a bundle for LPI, which no longer had to pay rent or keep a fundraising staff on its payroll.

Next, after a long and thorough search, Balz Frei was hired as director of LPI in the summer of 1997. The Institute spent the rest of the late 1990s setting up its research agenda and recruiting new faculty. In 1998 LPI hired Tory Hagen, Maret Traber, and Rod Dashwood, all acclaimed scientists whom Lawson described as the "research backbone of the Institute." Shortly afterward, David Williams was hired from within OSU as another principal investigator; he ended up holding numerous positions at LPI and was very important to its success in the following years.

In 2000 LPI launched one of its most successful projects: the Micronutrient Information Center. The resource, which continued to expand, provides information on dietary intake and encouragement for healthy living. While it still advises vitamin C doses much higher than those recommended by the U.S. Food and Drug Administration, the numbers involved are far from Pauling's recommended megadoses of the 1970s and '80s.

The Pauling centenary year of 2001 was another big one for the Institute, in part because of the first Diet and Optimum Health Conference held that winter. As part of the conference, the inaugural Linus Pauling Institute Prize for Health Research was given to Dr. Bruce Ames, along with a $50,000 award. In the span of a decade, the Institute had gone from being hundreds of thousands of dollars in debt to being able to award a biennial prize of $50,000 — tangible evidence of a truly remarkable turnaround. That year, LPI also hired Joe Beckman, who opened up a new area of research for the Institute through his focus on amyotrophic lateral sclerosis, or Lou Gehrig's disease.

The ensuing decade was refreshingly free of drama — certainly so by past LPISM standards — and saw unprecedented growth. In 2002 the general expansion of LPI's research support staff continued and in 2003 the second Diet and Optimum Health Conference was held with the signature prize going to Dr. Walter Willett of Harvard University. The third, fourth, and fifth conferences were held in 2005, 2007, and 2009 with Drs. Paul Talalay, Mark Levine, and Michael Holick winning the awards at each event, and in 2011 the prize went to OSU alum Dr. Connie Weaver.

In an otherwise near-spotless decade of growth and good news, one tragic occurrence did befall LPI. On May 31, 2006, Jane Higdon, a prolific writer, well-known researcher, creator of the Micronutrient Information Center, and six-year veteran of LPI, was hit and killed by a logging truck while bicycling near her home in Eugene. In her honor, the Jane Higdon Foundation was established, the trucking company involved in the accident donated $1 million to bicycle safety programs, and LPI set up the Jane V. Higdon Memorial Fund. The Higdon Foundation's goal is to create "scholarships and grants to encourage and empower girls and young women to pursue healthy and active lifestyles and academic excellence" and also to promote bicycle safety in Oregon's Lane County. The Memorial Fund is largely dedicated to supporting the Micronutrient Information Center.

Buoyed by the success of its past outreach efforts, LPI decided to expand its education programs to include young people as well, launching the Healthy Youth Program in 2009. The Program targeted elementary- and middle school-age students, and promoted healthy lifestyles and nutrition.

At the same time, LPI responded to the "Physicians' Health Study II on Vitamin C and E and the Risk for Heart Disease and Cancer." Published in the *Journal of the American Medical Association*, the study claimed that vitamins C and E were ineffective in treating cardiovascular disease. LPI retorted that the research directly contradicted numerous other contemporary studies, that it failed to accurately measure vitamins in the bloodstream, and that a ten-year study wasn't adequate for gauging the effects of vitamins on cardiovascular disease. LPI's response was emblematic of its continuing willingness to contribute to conversations on, as Linus Pauling would have put it, "how to live longer and feel better."

For the first time in its existence, things were going very smoothly for LPI, and as the first decade of the new millennium came to a close, more exciting news was near on the horizon. A new chapter in Institute history got off to a great start when, in 2011, OSU opened the Linus Pauling Science Center to house LPI, parts of the Chemistry Department, and other lab and teaching spaces.

For the Institute, the historical importance of the completion of the Linus Pauling Science Center was difficult to overstate. The building, which is the largest academic facility on the OSU campus, was a serious undertaking — it cost $62.5 million to build the four-story, 105,000 square-foot research center. The funding was acquired through donations from the Wayne and Gladys Valley Foundation ($20 million), the Al and Pat Reser family ($10.65 million), 2,600 private individuals (~$600,000), and a matching bond ($31.25 million) from the state of Oregon. The facility was also a cornerstone achievement of The Campaign for OSU, which raised over a billion dollars to support university initiatives.

Constructing its own building on the OSU campus was a goal for LPI from the minute the Institute moved to Corvallis. Indeed, Linus Pauling Jr. remembers sketching potential plans on napkins while at meetings with OSU staff during the moving process, and director (1997–2016) Balz Frei later wrote that, ever since LPI moved to OSU, building "a state-of-the-art research facility to house the Institute and serve as a high-profile working memorial for Linus Pauling" had been one of LPI's highest priorities.

The Linus Pauling Science Center was opened on October 14, 2011. Over 250 people attended the ceremony, during which Linus Pauling Jr. and OSU President Ed Ray delivered the main speeches. In his remarks, President Ray noted his belief that "preventive health care is the future of medicine," and that LPI and the Linus Pauling Science Center are in strong positions to develop this in the 21st century.

The center was designed by the firm ZGF Architects LLP, based in Portland, Oregon. It is a unique building with large windows and ample natural light. In addition, each of its floors is home to several works of art — including multiple light paintings created by Massachusetts-based artist Stephen Knapp — and those who work in the facility enjoy an enviable lunch spot on a fourth-floor balcony looking toward the Coast Range mountains.

Its lab space, however, is the real highlight of the Linus Pauling Science Center. Unlike most facilities, its labs consist mostly of open space, with only a few partial walls separating research areas. In a newspaper interview, Steve Lawson commented on this decision, noting, "We didn't want a lab environment with a lot of walls... For us, it's a way to keep the Institute coherent and increase the possibility of people communicating." In further pursuit of this goal, most of the Institute's noisier lab equipment is kept behind closed doors in dedicated spaces away from the work environment, thus rendering the laboratories a more pleasant place to think and interact. Seventeen years after moving to Oregon with a core staff of five, LPI had regenerated its roster to 63 employees and was home to a thriving research enterprise led by world-class faculty.

*

Linus Pauling Science Center grand opening Keynote Address, by Linus Pauling Jr., MD, October 14, 2011.

Linus Pauling Jr. at the dedication of the Linus Pauling Science Center, Oregon State University.

This is a very personal account of the background that has miraculously led to this wonderful, beautiful and exciting building, I title it: OUT OF ASHES THE PHOENIX ROSE.

It was back in the spring of 1991, just over 20 years ago now, that I sat down to talk with my father at his Big Sur ranch on the rugged

California coast. For many years, in fact since my mother died a decade earlier, my wife and I had made a pilgrimage to the ranch to be with my father and celebrate our three birthdays, which fortuitously fell within a two-week period.

I had been on the board since the Palo Alto Linus Pauling Institute of Science and Medicine's inception in 1973, so at our 1991 meeting I knew the situation had become desperate. My father, who for all his earlier life had been full of remarkable energy and ambition, now at 90 had lost that energy and was making mistakes in judgment. He was ill with the cancer that would kill him three years later.

LPISM was failing: half a million dollars of debt, laboratory research had vanished for lack of incentive and direction, donor income was being diverted to non-nutritional investigations, there were no research grants and morale was in the basement.

As his oldest son, I could not just stand by and watch this great man's efforts of the past quarter-century go down the drain, along with his reputation. If the Institute failed, all the naysayers would crow and describe him as a senile crackpot in spite of his astonishing lifetime achievements. Additionally, the thousands of donors over the years and the makers of future bequests would feel betrayed. It was obvious he needed help. As his son, I felt it was necessary to provide that help and it felt good to me to try.

So we had to talk. Early in my life I realized that my father was a very special person with talents I could never hope to emulate. That was emphasized by this story which I enjoy telling. When I was about 15, my father was writing an introductory chemistry textbook for Caltech freshmen, the best and brightest college freshmen, the cream of the crop. At the end of each chapter were questions. He asked me to read a chapter and answer the questions. I tried, valiantly, but I did not understand the text and could not answer a single question. When my mother heard about this, she hurried down to the Pasadena City Hall to have my name officially changed from Linus Carl Pauling to Linus Carl Pauling Jr. so no one could possibly mistake me for him.

At least I had sense enough to follow a very different track from my father, one that eventually gave me skills that now could be used to help him as my thanks to him for bringing me into the world.

It was now or never, so I boldly waded in. He and I discussed the future, starting with the past. I talked about his amazing life with his multiple triumphs in so many and so very diverse arenas.

His fame was worldwide, originating with the scientific community. I pointed out that he was arguably the first, and certainly the most successful, bridge-builder between chemistry, mathematics, physics, medicine and biology, linking these disciplines to create what is now the most popular science of all, molecular biology. One result of his creativity, hard work and dedication to science, as you all know, was the Nobel Prize for Chemistry.

It was during this time period that his interest in nutrition originated, spurred by his own life-threatening kidney disease. Thanks to a rigid diet prescribed by Stanford Medical School nephrologist Dr. Thomas Addis at a time long before renal dialysis, and carefully supervised by my mother, my father not only survived a usually fatal disease but recovered completely.

After World War II, prompted by my politically liberal mother whom he certainly loved deeply and wanted to please, he embarked on a spectacularly successful two decades of humanitarian effort, educating the governments of the world and, necessarily, their peoples, about the evils of war and the dangers associated with unrestricted exposure to radiation, especially that produced by the hundreds of nuclear bomb tests being conducted. He suffered vilification by many from all parts of the world. He was hounded by the FBI and the United States government.

His crowning moment of glory, at least in my estimation, was his indomitable courage in confronting those nasty witch hunters, the United States Senate Internal Security Subcommittee, when facing imprisonment when he refused to disclose the names of his ban-the-bomb United Nations petition assistants. He knew that these conscientious people, most of them scientists, would be less able than he was to defend them-selves from accusations and loss of employment. The Subcommittee, when faced by my father's public popularity, courage, remarkable mem-ory and command of facts, then backed off, their collective tail between their legs. His worldwide influence was so extensive and the result so positive that he was eventually awarded the Nobel Peace Prize.

So what was next for him? His old interest in nutrition as a factor in health and well-being resurfaced. Starting with vitamin C, he promoted nutrient research and encountered resistance from university, medical and government bureaucracies. He turned to the public, writing article after article and giving hundreds of talks, with the result of an explosion in popular food supplement usage. But research remained a fundamental necessity, so the private non-profit Linus Pauling Institute of Science and Medicine was founded in 1973 and initially showed promise.

By the time of our talk in 1991, LPISM's outlook was dismal.

At age 90, my father was tired and dispirited. Being fully occupied with his own illness, he was unwilling to devote energy to coping with his Institute's problems. I said to him that I could not in good conscience stand by and see his eponymous Institute go down in ignominious defeat. With his incredibly illustrious past, I felt strongly that he deserved more than that. And maybe, just maybe, I could do something about it.

We decided, together, that if the Institute, and also his reputation, were to survive, the best course of action was for the Institute to affiliate with a reputable university. That would ensure the rigorous scientific attitude and protocol necessary to legitimize micronutrient research in the future. And, most important of all, we had to be ethically responsible to the thousands of past, present and future donors who believed in my father and supported the Institute. We could not let them down.

I had just retired from 35 years of the practice of psychiatry, so I had the time and energy to devote to other endeavors. After discussion with my wife, I decided to offer to take over management of the Institute. I had to have my wife's agreement, because I was planning to spend considerable time in Palo Alto, a long way from my home in Honolulu.

To his credit and with an audible sigh of relief, my father agreed. We discussed affiliation possibilities, Stanford and Caltech among them. He seemed, however, to favor Oregon State University, his undergraduate alma mater, to which he had already committed his scientific papers. If you haven't already, you should check out the Pauling Papers at the OSU Valley Library Special Collections website. You will be impressed.

During the next years, I became President and Chairman of the Board of LPISM. We reorganized radically and survived many trials and tribulations. My essential second-in-command Steve Lawson and I visited many universities.

OSU, thanks to then President John Byrne, Development Director John Evey and Dean of Research Dick Scanlan, was our clear and undisputed choice.

And what a great choice it was! Here now, before us, 15 years later, is the Linus Pauling Science Center, dedicated to highest-quality research in scientific areas that would surely be of interest to my father. I'm sure, if he were here, he would have tears of joy in his eyes just as I do.

I want to thank OSU President Ed Ray, Dean Sherman Bloomer, LPI Director Balz Frei, architect Joe Collins, the many others in the system who have participated in making this possible, all the donors and the people of the great state of Oregon. I specifically thank the key major donors, Tammy Valley and Pat Reser, for allowing Linus Pauling's name to be on this beautiful building. That is a very unusual act of generosity.

It will be a great future. Thank you all with my whole heart.

21 A Funny Story from Andy Warhol

Chapter

By Chris Petersen

Linus Pauling, Jill Sackler, and Andy Warhol in New York, November 21, 1985. The Linus Pauling Medal for Humanitarianism was presented to Arthur Sackler that evening.

As recorded by Warhol in his diary:
Thursday, November 21, 1985

"...the Sacklers were doing this thing at the Metropolitan Club and I was figuring out who to bring, and I should have brought Dr. Li, I guess, because I wound up sitting with Dr. Linus Pauling, but I brought Paige and she had a really good time. [...] So I cabbed to the Metropolitan Club ($5). And there's Paige sitting downstairs in the hallway.

Those horrible doormen there wouldn't let her in because she didn't have a fur coat! [...]

"And Dr. Pauling took my arm, he was getting an award. Upstairs I was next to Jill Sackler, across from Martha Graham, and Jill said, 'Martha's been dying to meet Linus Pauling for years and now she's next to him and doesn't know it.

"I met a man who said he invented vitamin B or C. And Dr. Pauling was telling us that the only real killer is sugar, and then Paige and I were dumbfounded later when they brought dessert and he sat there eating all these cookies."

This brings to mind a favorite Pauling anecdote of ours, which he told in 1987:

"Usually I eat two eggs in the morning, sometimes bacon, but I happen to be lazy enough not to cook more than one thing for a meal. The last two days I was eating oxtail soup with vegetables. I don't know what I'll have today. Perhaps some fish. In my book [*How to Live Longer and Feel Better*] I say you shouldn't eat sweet desserts, but I also quote a professor who says that this doesn't mean that if your hostess has made this wonderful dessert you should turn it down. My wife used to say I always looked for that hostess."

22 Quasicrystals

Chapter

By Jaren Provo

Pauling speaking in 1986.

In 2011 Dan Shechtman, a distinguished professor at the Israel Institute of Technology (Technion), was awarded the Nobel Chemistry Prize for his discovery of something called quasicrystals, a collection of atomic structures whose order defies traditional crystallographic understanding. Though eventually recognized with the highest decoration that a chemist can receive, Shechtman's discovery shocked the condensed matter science community when he and his team published their breakthrough in 1984. The strange qualities that quasicrystals exhibited —

317

including "five-fold symmetry," which had previously been dismissed as impossible — led to extensive research, debate, and even disbelief.

Among the scientists engaged in this debate was Linus Pauling, who studied quasicrystals intermittently between 1985 and 1993. Pauling's own deep background in crystallography made him resistant to the discovery's implications: that the discipline's core principles about what was possible in solid structures were insufficient. In response to Shechtman's ideas, Pauling instead proposed a "multiple-twinning" theory that used, rather than contradicted, traditional crystallography's assumptions to explain the unusual qualities of quasicrystals.

More specifically, Pauling's thinking was that quasicrystals were not a new form of material existing somewhere between crystalline and amorphous solids, but that they were really twinned crystals, made from the same unit cells seen in crystalline materials. The substructures from which the unit cells were made, he claimed, were oriented differently in space and embedded in one another as twins. The five-fold symmetry that appeared, Pauling argued, was the result of multiple twins and was not found in the single substructures themselves. This given, the "forbidden" five-fold symmetry would not actually be "true" symmetry, because it would not all happen in one spatial orientation (also called a "domain"). Instead, it would be a "false" symmetry made by the multiple twins.

Because Pauling believed quasicrystals were formed from multiply twinned crystals, and not single crystals, he decided to rely on finely crushed powder samples and X-ray beams to conduct his analysis. By doing so, the patterns that resulted would not be influenced by twinning, since the domains of the twins would be randomly oriented. The trade-off, however, was that the powder patterns were harder to work with, since they produced rings, rather than spots, making the symmetry unclear. This meant that, to form a model for his twinning theory, Pauling had to rely on ring measurements that were often difficult to determine precisely.

Despite this limitation, the model that Pauling developed was intricate, complex, and impressive in its spatial reasoning. But though it initially seemed like a promising hypothesis, it was unsupported by evidence and, as we will see, was gradually abandoned by the condensed matter community.

*

In part because of the presumed impossibility of five-fold rotational symmetry, Shechtman was shocked when he had found an alloy ($MnAl_6$, containing aluminum and manganese) that exhibited these traits. Shechtman elaborated on his surprise in a 1985 letter to Pauling, in which he admitted his initial disbelief and detailed at least four other kinds of experiments to which he had subjected his findings, in addition to asking other researchers to review and duplicate his results.

The reviewers agreed: the alloy Shechtman studied had five-fold rotational symmetry as well as a related spatial arrangement called icosahedral symmetry. The alloy also exhibited certain kinds of order similar to crystals, but not translational symmetry, a simple form of arrangement that can be used to easily define a crystal. Shechtman's study, "Metallic Phase with Long-Range Orientational Order and No Translational Symmetry," was published in 1984 in the major peer-reviewed journal *Physical Review Letters*, and the name "quasicrystals" emerged soon thereafter.

Pauling's counterproposal to quasicrystals was known as the "multiple-twinning" hypothesis. Prior to the discovery of quasicrystals, crystallography held that some structures exhibited a phenomenon called "twinning." In twinning, crystals with the same structure exist in different domains — that is, they are oriented so they are essentially facing in different directions — but are embedded within each other, effectively making a new structure altogether. One way to visualize twinning is to imagine crystals as being formed from "clusters" of small sets of atoms, with some of the clusters sharing their "end atoms" such that two clusters stem from a shared set. These clusters are thus "twinned."

Pauling felt certain that quasicrystalline structures could be explained by multiple-twinning between atomic clusters in the crystal. Analyzing Shechtman's article, he asserted that a large, roughly cubic unit cell with twinning clusters was responsible for the apparent icosahedral symmetry.

To aid in developing his multiple-twinning hypothesis and to test some initial predictions, Pauling approached David and Clara Shoemaker, a husband-and-wife team of crystallographers working at Oregon State University (OSU). David had previously collaborated with Pauling on X-ray diffraction while studying under him as a graduate student at Caltech, and in a speech given in 1995 at the OSU symposium, "Life and Work of Linus Pauling: A Discourse on the Art of Biography," he recalled Pauling insisting that, contrary to Shechtman's claim, the $MnAl_6$ structures he had found could be indexed to a Bravais lattice (a three-dimensional concept commonly used in conventional crystallography), albeit through a complex interchange of twins. Shoemaker's reflections reinforce the idea that, above all, Pauling was certain that the rules of crystallography did not need to be modified to accommodate quasicrystals.

Pauling felt experimental data substantiated his twinning model for a variety of reasons. First, his initial calculations for the unit cell size — approximately 26.73 Å — matched X-ray powder images given to him by Shechtman. Second, Pauling had found what he called faint "layer" lines in the powder images that he felt were not adequately explained by quasicrystal theory, but instead matched structures with multiple twins. Third, Pauling noted that the most pronounced spikes ("Bragg peaks") in the powder images were shifted from their expected locations in ways that could be accounted for by his twinning model, but could not be addressed by the model for quasicrystal growth; that is, some atoms were in unexpected positions that could not yet be explained by any other theory of how quasicrystals arranged themselves. Most of all, Pauling's repeated insistence on his experience with and integral role in shaping crystallography signaled his resistance to changes in foundational concepts. For Pauling, the existing tenets of crystallography remained sound enough to explain what others were quick to call an exception.

But Pauling's twinning model had significant problems. David Shoemaker recalled having initial success with the X-ray diffraction patterns, finding that they matched Pauling's calculations for the unit cell side length, at 26.73 Å. But when Pauling revisited his calculations to confirm their accuracy, the work hit a snag. Instead of a 26.73 Å unit cell side, Pauling realized his calculations called for a 23.36 Å side, a difference of about 15%. From Shoemaker's perspective, this made the theory implausible. "I don't think he was successful," Shoemaker stated with respect to Pauling's argument. "We [David and Clara] examined the figures ourselves and were unable to find any justification for the twinning theory there. So we, perhaps understandably, lost interest in it, but he continued on."

Indeed, Pauling began publicly arguing for multiple-twinning in late 1985. In a *Nature* interview with John Maddox, he introduced his ideas to a broader audience, showcasing a 1,120-atom unit cell to describe the $MnAl_{12}$ structure. Having done so, he concluded that "Crystallographers can now cease to worry that the validity of one of the accepted bases of their science has been questioned." Shortly afterward, Pauling submitted a letter to the magazine *Science News*, which the periodical titled "The nonsense about quasicrystals." In it, Pauling wrote:

"There is no doubt in my mind that my explanation of the quasicrystal phenomenon is correct. I have now accounted for the atomic arrangement seen on the electron micrographs. I trust that my paper containing these additional arguments will be published in *Physical Review Letters*. I think that it is interesting that an intermetallic compound that I investigated in 1922, and whose structure was determined 40 years later, has the same structure as these 'quasicrystals,' but without the twinning that they show. This is the compound sodium dicadmide, which is mentioned in my *Nature* article. It is also interesting that the scientific journals are printing scores of papers about exotic explanations of the observation but that I have had difficulty getting my papers on the subject published. I think that I am almost the only, perhaps really the only, X-ray crystallographer who has become interested in this subject. The explanation probably is that the other X-ray crystallographers felt that the nonsense about quasicrystals would soon fade away. That is how I felt for about five months, and then I finally decided that I would look into the matter."

The letter was published January 4, 1986, only three months after the publication of Pauling's first article on the subject (the "*Nature*" article" he mentions). His claim that no other X-ray crystallographers were interested in quasicrystals was an exaggeration, but the bulk of the scientists concerned with the subject were, in fact, physicists and not analytical chemists.

That some referees at *Physical Review Letters* allegedly felt Pauling was behaving as "an antagonist" toward quasicrystal theorists — and perhaps Shechtman in particular — is not surprising, given the tone of his *Science News* piece. Describing the discovery and related research as "nonsense," saying that X-ray crystallographers avoided the matter and hoped it would "fade away," and referring to initial explanations by other scientists as being "exotic" are indicative of Pauling's condescending attitude toward the work. Further, Pauling's mention of his own extensive expertise in the field of crystallography, coupled with his seemingly patronizing line, "I finally decided I would look into the matter" tempts one to conclude that Pauling believed, perhaps a bit too strongly, in his own superiority.

*

The Shoemakers were not the only scientists who felt that Linus Pauling's quasicrystals hypothesis, while novel, was unsubstantiated by experimental data. In fact, in "Metallic Phase with Long-Range Orientational Order and No Translational Symmetry," the article that started it all, Shechtman and co-author Ilan Blech noted that twins were initially suspected as being the source of the unusual structure, but that after subjecting the crystals to a broad range of experiments and even using data from the X-ray diffraction patterns themselves, they determined that their sample was not composed of twins.

The authors' argument centered on the fact that twins should have been visible when examined through a method called "dark-field microscopy," which illuminates whole grains, and that the twins should also have changed, in some fashion, the patterns produced by their electron diffraction experiments at various resolutions. They further suggested that twinning would not interfere with matching a Bravais lattice to a crystal in an X-ray diffraction pattern. Therefore, the fact that scientists were having difficulty assigning lattices — and, by extension, unit cells — to quasicrystals would not, in itself, be an indication that their structure was based on twins. Because of this overwhelming evidence against, the research team concluded that the sample "[did] not consist of multiply twinned regular crystal structures."

But Pauling persisted in his line of argument. In a letter dated April 24, 1985, he mentioned that certain previously analyzed crystalline structures shared some of the characteristics of the newly discovered quasicrystals, and that there had been structures in the past that initially surprised researchers, but were eventually found to accord with existing paradigms in the field. He concluded by requesting copies of Shechtman and Blech's $MnAl_6$ X-ray diffraction data, so that he might continue developing his theory.

Shechtman, working in Haifa, Israel, responded promptly. In a letter received by the Linus Pauling Institute on May 15, he emphasized that a number of different experiments had yielded no evidence of twins, and that other research teams had backed his findings. Nonetheless, Shechtman included prints of the X-ray diffraction patterns and copies of the team's Bragg peaks data.

In a response dated June 6, Pauling thanked Shechtman for the diffraction images, and requested permission to use them in an article introducing his multiple-twinning theory. He explained, very briefly, his hypothetical twinned, 20-icosahedral structure as the basis for the icosahedral symmetry seen in quasicrystals, and mentioned certain pieces of evidence that he felt Shechtman and his team had overlooked. In particular, Pauling noticed three to five weak lines in the X-ray diffraction patterns that, for him, supported a twinned structure.

Then Pauling made an unexpected offer: he asked Shechtman if he would like to co-author the twinning article with him. The difficulty, Pauling pointed out, would be that Shechtman would have to contradict himself and, to some extent, negate his own findings. In essence, Pauling offered Shechtman an opportunity to admit that he was wrong. It is unclear what, exactly, Shechtman's response was; no letter from him about the offer is preserved in the Pauling Papers, and the tone of Pauling's next letter implies that the two did not discuss it.

In mid-July 1985, Pauling again wrote to Shechtman, revealing a proposed structure composed of "icosatwins" and 1,000-atom unit cells. He acknowledged that the unit cell size was unusually large, but stood by his hypothesis, adding that he himself had discovered a few complex crystal structures during the 1920s, an era when 1,200-atom cells were inconceivable. He also mentioned that, since he had not received a reply from Shechtman indicating interest in co-authorship, he had sent the article off to *Nature* for review. (The article would be accepted and published later that year.)

Records indicate that, shortly after this exchange, Shechtman and Pauling met in person at the Pauling Institute in Palo Alto. Mentioning that he would be in America on business, and available during the week of August 19, 1985, Shechtman wrote that he hoped they could meet and discuss quasicrystals in detail. Pauling obliged and extended an invitation, and though the particulars of the meeting are unclear, it does not seem to have gone especially well.

In a letter dated September 3, 1985, Shechtman wrote to Sten Samson, a Caltech faculty member and former Pauling graduate student, carbon-copying Pauling, and enclosing a sample of a mostly — but not entirely — icosahedral, rapidly cooled, powdered ribbon of material. He wrote that he was fulfilling a request made by Pauling, and that Samson had, apparently, been forewarned that he would be receiving the sample and was aware of the analyses that Pauling wanted performed. The tone of Shechtman's letter indicates a measure of strain in the relationship between the two principals involved.

The results of Samson's analysis are not extant among the Pauling Papers, but correspondence between Pauling and Shechtman appears to have dissipated for nearly a year. In mid-April 1986, Shechtman sent a brief letter requesting that Pauling keep him apprised of his activity with regard to quasicrystals. No reply on Pauling's part is preserved amongst his papers.

It seems it was not until August 12, 1986 that Pauling again wrote to Shechtman. Addressing both Shechtman and Ilan Blech, Pauling claimed that he had found an error of 14.4% in their scale for the electron diffraction patterns of $MnAl_6$ — the sample used in their first article — and that, adjusted against this error, the data strongly supported Pauling's twinning hypothesis. He went on to call the error "easily avoidable" and (in a somewhat patronizing tone) detailed a fairly basic process by which the team could have — but, Pauling presumed, did not — verify their scale's accuracy. Pauling's tone was curt: "This error caused me several months of unnecessary effort," he wrote. He went on to say that, after correcting this presumed mistake, "there no longer remains any doubt about the nature of the 'icosahedral quasicrystals.' They are twins of a cubic crystal with edge 26.73 Å." Pauling closed his letter with a request of sorts: "Verification of my statement that your scale is in error by 14.4% would, of course, provide additional evidence for the foregoing conclusion."

Though Shechtman's response is not among Pauling's correspondence, it seems to have had a humbling effect. A letter dated September 8, 1986 — written less than a month after Pauling's terse note — reveals Pauling to be in a more deferential mood. In it he thanked Shechtman for his letter and photographs, and acknowledged that they clearly showed that Shechtman and Blech's original quasicrystals scale was correct. The error, Pauling wrote, occurred from misinterpreting what he called a "statement about scale made to me by another investigator."

Despite acknowledging his blunder, Pauling did not directly apologize, but instead went on to state concerns about a different set of photographs showing decagonal relationships, estimating calculations based on them to be off by 8%. The ensuing gap in correspondence in the Pauling Papers implies that Pauling and Shechtman did not communicate again for an entire year.

However sparse the exchange between Shechtman and Pauling during that time, it did not mirror a reduction in Pauling's work on his twinning hypothesis. The news that Shechtman and Blech's scale was indeed accurate caused a small crisis for Pauling, who was certain that the "correction" he proposed would justify his structure. For much of October 1986, Pauling contemplated the multiple-twinning structure at either his ranch in Big Sur or on airplanes and in hotel rooms between peace talks and chemistry lectures. Dozens of pages of calculations, diagrams, and hypotheses on legal pads reveal a constant refinement of the twinning theory. Pauling's meticulousness in noting the date and time reveals that the "quasicrystal problem," as he called it, occupied his mind even into the early morning hours.

On October 16, 1986, Pauling arrived at a 920-atom unit cell composed of eight 117-atom clusters arranged snugly at 90-degree angles of rotation, repeating along all three axes. Triumphantly, Pauling wrote "Hurray! The Quasi Problem is Solved" in his manuscript. But a small note in the margin underneath redirects the reader to work done two days later, in which Pauling recalculated his cell size based on Shechtman and Blech's electron diffraction photos. The new cell was still composed of eight 117-atom clusters, but as many as 72 of those atoms were shared, making the unit cell 840 atoms instead of 920. Pauling also concluded that the cell would be essentially body-centered cubic. Eventually that cell also succumbed to scrutiny, and on October 24, Pauling considered an 804-atom unit cell.

Finally, Pauling concluded that the unit cell likely contained 820 atoms, formed from 104-atom clusters sharing outer electron shells. This structure was the basis of Pauling's article, "Evidence from X-ray and neutron powder diffraction patterns that the so-called icosahedral and decagonal quasicrystals of $MnAl_6$ and other alloys are twinned cubic crystals," published in June 1987 in the *Proceedings of the National Academy of Science (PNAS)*.

Pauling's clusters were arranged such that each was at the corner of a unit cube and surrounded by 12 more clusters in the shape of a nearly regular icosahedron. "All of the clusters have the same orientation," he put forth, "and any one cluster could serve as a seed for twinning." To defend this structure, Pauling pointed to its considerable correlation with diffraction patterns, and argued that any mismatching that occurred around cube edges was due to slight variations in alloy compositions. He also noted that another alloy, $Mg_{32}(Al_9Zn)_{49}$, is known to have a tightly packed cluster-based structure, making such a structure in $MnAl_6$ not unprecedented. In fact, Pauling argued, the intense heating and rapid cooling process used to form $MnAl_6$ crystals (which are typically referred to as "rapidly quenched") likely led to closely packed alloy clusters. By the time of the article's publication, over 100 other alloys with icosahedral quasicrystalline structures had been found. Pauling, like many scientists, began to expand his theory to account for even more alloy structures, beyond the first anomalous discovery, $MnAl_6$.

Breaking the apparent year of silence, Pauling wrote once more to Shechtman on October 6, 1987. By now his tone had shifted from cordially tense scientific competition to camaraderie. In addition to gratitude for the many glossy X-ray photographs and diffraction data calculations he had provided him over the years, Pauling also thanked Shechtman for his very discovery of quasicrystals. He wrote,

> "This discovery has resulted in a great contribution to crystallography and metallurgy, in that it has stimulated hundreds of investigators to study alloys and has led to much additional knowledge about intermetallic compounds... Your discovery has also made me happier... For over two years I have worked on this problem, and have enjoyed myself while doing it. I estimate that I have spent nearly 1,000 hours

just thinking about this whole question, and more than 1,000 hours making calculations, and writing papers."

The fruit of this labor, for Pauling, was the discovery of "five new complicated structures," the details of which he shared with Shechtman. In what looks like a conciliatory acknowledgement of expertise, Pauling even asked Shechtman to check on the diffraction patterns of a slowly cooling structure to see if the intensity spots shifted in position or intensity.

Pauling then reiterated his previous desire to write a paper with Shechtman about two- and six-fold symmetry, confiding, "I should be very pleased if a paper could be published with the authors Shechtman and Pauling." From there he acknowledged, to some extent, the tension that had existed between them, writing, "I hope that we can cooperate in the attack on this problem. I have the impression from referees' reports on papers that I have submitted to *Physical Review Letters* that at least one of the referees considers me to be an antagonist." Though Pauling did not agree with this assessment, or apologize for any of his behavior, he did write, "It is something like the situation between the United States and the Soviet Union. It would be much better if they were to cooperate in attacking world problems, rather than to function as antagonists."

Shechtman promptly cabled a reply, thanking Pauling and saying that his letter "made me very happy." In a follow-up sent on November 10, Shechtman expressed interest in collaborating with Pauling on a joint quasiperiodic structures article, and offered to host Pauling at Technion in Haifa, covering all of his expenses. In response, Pauling wrote that he was cutting down on his travel, and so would not likely make it to Israel. Instead, Pauling wrote, perhaps he would send Shechtman manuscripts for his consideration and input as occasion arose. That letter, dated December 8, 1987, is the last archived bit of correspondence between Pauling and Shechtman. The two never co-authored an article.

*

Linus Pauling was not the only scientist to offer an alternative theory for the nature of quasicrystals. In fact, one of the major competing theories, the "icosahedral glass" theory, was actually introduced and then quickly abandoned by Shechtman and Blech themselves, before further development by physicists Peter Stephens and Alan Goldman. More importantly though, Shechtman was not the only scientist who held fast to and developed quasicrystal theory; over time, a growing number of physicists and crystallographers began to support the idea that quasicrystals were legitimate exceptions that warranted redefining what qualified as a crystal.

In the November 1989 issue of *PNAS*, an article written by P.A. Bancel, P.A. Heiney, P.M. Horn, and P.H. Steinhardt, and titled "Comment on a Paper by Linus Pauling," addressed Pauling's ever-evolving multiple-twinning hypothesis. The paper responded in particular to his article, "So-called icosahedral and decagonal quasicrystals are twins of an 820-atom cubic crystal," also published in *PNAS*. Prior to the publication of Pauling's "Icosahedral quasicrystals of intermetallic compounds are icosahedral twins of cubic crystals of three kinds," the team sent their comment to Pauling for his consideration. Pauling encouraged them not only to publish the article, but to publish it simultaneously with his own, so that they would appear in the same issue of *PNAS*. In fact, Pauling himself communicated their finished manuscript to the journal.

Just before submitting their "Comment on a Paper by Linus Pauling," Bancel and his co-authors formed a sample of AlFeCu alloy that was considered "perfect," in that the crystal creation process was refined to produce extremely few anomalies and deformations. They then examined their new sample from the perspective of Pauling's twinning hypothesis, and concluded that Pauling would need to employ a unit cell containing nearly 100,000 atoms to describe an imperfect sample of AlFeCu alloy, and one of over 425,000 atoms to account for the team's "perfect" samples. Such a structure, they argued, would be unfathomably complex, and an impractical model of the material's structure.

According to the team, the apparent success of Pauling's hypothesis in part owed to the presence of structures called "phason strains."

To understand the idea of a phason strain, one must first know that stresses applied to crystalline structures cause deformation, and that a variety of imaginary particles, called quasiparticles, have been devised for the sole purpose of describing how physical reactions change the nature of certain subatomic particles. Instead of explaining, for example, how an electron's behavior is modified by its interactions with other electrons in surrounding atoms — a dauntingly complex task — one can simply substitute a particle that resembles an electron, but is more massive. Such an imaginary particle behaves quite similarly to the electron in its interactions, but requires less complicated modeling, as it is essentially standing in for the behavior of a whole group of interacting particles.

Two quasiparticles pertinent to crystallography are phonons and phasons. When external stress is applied to a crystalline structure, unit cells of that structure are distorted from their equilibrium shapes. This distortion is referred to as a "phonon strain." When the stress is released, the return to equilibrium shape is modeled as the strain "relaxing" by transmitting phonon quasiparticles at the speed of sound. (Bear in mind that phonons do not really exist as particles, but are being employed for the sake of simplifying the model.) Effectively, the crystal structure returns to its original state immediately.

However, applied external stress can have another kind of effect on a crystalline structure. Instead of distorting the overall unit cell shape, stress may rearrange unit cells without appreciable shape distortion. In "proper" crystals, described by uniform unit cells, such distortion would have no noticeable effect; the roughly identical parts would only be shuffled around, and the overall structure would look essentially identical. In contrast, the non-periodic structure of quasicrystals — their lack of translational symmetry — means that rearranging parts of the overall pattern would change the structure noticeably. This rearrangement is referred to as a "phason strain." Unlike phonon strains, they do not relax instantly once external stresses have been released. Instead, the process of returning to the ground state configuration may take hours, days, or even years. Thus, the shifted structure remains long after an explanation for the modification is visible.

In quasicrystals, phason strains break icosahedral symmetry and change the ratio of distances between the structural parts, such that it is no longer a fixed irrational number. This has the effect of shifting diffraction peaks from their expected locations and distorting the regularity of the X-ray patterns. When such distortions are visible, the two logical conclusions from these shifts are that the crystal either has frozen-in phason strains or is formed from a very large, twinned unit cell.

For Bancel and his team, the reason why Pauling's twinning model appeared to match experimental diffraction data was because the unit cell it arrived at for each compound was comprised of the atoms between phason strains, which appeared to act as the boundaries to large, distinct unit cells. Twinning theory, they pointed out, also has the virtue of responding more directly to diffraction peak shifts, since it fits a unit cell specifically to the deformations that a sample exhibits.

But the fundamental difficulty with Pauling's model was that the data simply did not provide true evidence of twins; in particular, certain artifacts that result from twinning were not present as expected in the data. The immense unit cells the theory required also made for impractical models that could not be generalized, especially given that they had to become larger and more complex as the quasicrystal samples being modeled neared perfection. In fact, assuming every new sample of the same alloy has different diffraction peak shifts — a reasonable assumption, given that Bancel, et al. demonstrated that quasicrystals can be refined to eliminate peak shifts almost entirely — the multiple-twinning hypothesis would technically require that a new unit cell be devised for each new sample, each custom tailored to its unique circumstances.

The Bancel group concluded that Pauling's multiple-twinning hypothesis was inadequate. Instead, by acknowledging that quasicrystals are a legitimate exception to traditional tenets of crystallography — albeit ones requiring restructuring of the definition of a crystal — it becomes possible to model the phenomenon more simply and accurately, and is therefore a better explanation than multiple-twinning.

In private correspondence, a member of the Bancel group, University of Pennsylvania physicist Paul Steinhardt, wrote to Pauling reiterating the importance of his team's "perfect" crystals and that their implications would "place very severe constraints on any multiple-twinning model." He implied that Pauling had both objected to Bancel's stance and hypothesized that the diffraction patterns Bancel derived from his sample only matched theoretical values because of a phenomenon called "multiple-scattering." However, Steinhardt noted that Bancel had, as a follow-up to an exchange between Steinhardt and Pauling on the subject, done some "sample rotation experiments" to confirm that the diffraction data did, in fact, support the team's claims.

Undaunted, Pauling continued to work on his multiple-twinning hypothesis, refining and applying the model to a variety of alloys. In 1993 he received a letter from Dr. Simon C. Moss, a physicist at the University of Houston, in which the author expressed his belief that electron microscopy, diffraction evidence, and the twinning theory's "absurdly large approximant cells" had all effectively ruled out Pauling's model. Moss did concede that it could be possible that quasicrystalline forms may not be in their ground states, and that they may form "multi-domained complex crystals" (that is, twinned structures) at lower temperatures, offering a small concession to twinning theory's potential. But of quasicrystal theory itself, Moss wrote, "We will certainly keep you informed on our progress and perhaps, in time, bring you to our point of view. It is, I should say, rather widely held and well-supported by the data."

Regardless, there is no indication that Moss, nor the growing number of chemists and physicists supportive of quasicrystal theory, succeeded in swaying Pauling. In his written response to Moss, Pauling pointed out a variety of small "horizontal and vertical layer lines" visible in overexposed photographs, which he felt were inadequately described by quasicrystal theory. He also reiterated his belief that accrediting shifted diffraction peaks to the influence of phason strains was "unsatis-factory." Though he acknowledged his theory would require very large unit cells — 52 Å, 58 Å, and perhaps even 66 Å in width — he also pointed out that, 70 years prior, he had thought that a 30 Å structure with 1,000 atoms was overwhelmingly large, and that a structure of this sort was now accepted as accurate and reasonable by the scientific com-munity.

Pauling seems to have defended the multiple-twinning theory until his death in 1994, despite the growing evidence and support for the theory that quasicrystals were, in fact, anomalies that required the field to rethink what forms of ordered solids were truly possible. Today, in addition to their recognition by the Nobel committee, quasicrystals are finding potential use as insulation in engines, materials for converting heat into electrical energy, and components of wear-resistant ball bear-ing coatings, non-stick frying pan liners, LED components, and parts in surgical instruments.

23 Pauling's Last Years

By Matt McConnell

Pauling delivering his last lecture at the International Symposium on Biological NMR, Stanford University, March 25, 1994.

In 1917, at 16 years of age, Linus Pauling wrote in his personal diary that he was beginning a personal history. "My children and grandchildren will without doubt hear of the events in my life with the same relish with which I read the scattered fragments written by my granddad," he considered.

By the time of his death, some 77 years later, Pauling had more than fulfilled this prophecy. After an extraordinarily full life filled with political activism, scientific research, and persistent controversy, Pauling's achievements were remembered not only by his children, grandchildren and many friends, but also by an untold legion of people whom Pauling himself never met.

Passing away on August 19, 1994 at the age of 93, Pauling's name joined those of his wife and other family members at the Oswego Pioneer Cemetery in Oregon. What follows is an account of his final three years.

*

In 1991 Pauling first learned of the cancer that would ultimately claim his life. Having experiencing bouts of chronic intestinal pain, Pauling underwent a series of tests at Stanford Hospital that December. The diagnosis that he received was grim: he had cancer of the prostate, and the disease had spread to his rectum.

Between 1991 and 1992, Pauling underwent a series of surgeries, including the excision of a tumor by resection, a bilateral orchiectomy, and subsequent hormone treatments using a non-steroidal antiandrogen called flutamide. During this time, Pauling also self-treated his illness with megadoses of vitamin C, a protocol that he favored not only for its perceived orthomolecular benefits, but also as a more humane form of treatment than chemotherapy or radiation therapy.

Pauling's interest in nutrition dated to at least the early 1940s, when he had faced another life-threatening disease, this time a kidney affliction called glomerulonephritis (see Chapter 2). Absent the aid of contemporary treatments like renal dialysis — which was first put into use in 1943 — Pauling's survival hinged upon a rigid diet prescribed by Stanford Medical School nephrologist, Dr. Thomas Addis. At the time a radical approach to the treatment of this disease, Addis's prescription that Pauling minimize stress on his kidneys by limiting his protein and salt intake, while also increasing the amount of water that he drank, saved Pauling's life and led to his making a full recovery. Though his highly public interest in vitamin C would not emerge until a couple of decades later, Pauling's health scare instilled in him a belief that dietary control and optimal nutrition might effectively combat a myriad of diseases. This scientific mantra continued to guide Pauling's self-treatment of his cancer until the final months of his life.

Pauling also believed that using vitamin C as a treatment would, as opposed to chemotherapy, allow him to die with dignity. Were his condition terminal and his outlook essentially hopeless, Pauling adamantly felt that he should be permitted to pass on without "unnecessary suffering." Pauling's wife, Ava Helen, had died of cancer in December 1981. She too had refused chemotherapy and other conventional approaches for much of her illness, a time period during which her husband had helped in the best way he knew how: by administering a treatment involving megadoses of vitamin C. This attempt ultimately failed and, by his own admission, Pauling never really recovered from Ava Helen's passing.

Nonetheless, in his capacity as researcher and administrator at the Linus Pauling Institute of Science and Medicine (LPISM), Pauling continued to lead research efforts to substantiate the value of vitamin C as a preventative for cancer and heart disease. By the time of his own diagnosis in 1991 however, the Institute was in a desperate financial situation, deeply in debt and lacking the funds necessary to pay its staff.

In 1992, while he recovered from his surgeries and managed his illness, Pauling continued to act as chairman of the board of LPISM. No longer able to live entirely on his own, he split his time between his son Crellin's home in Portola Valley, California, and his beloved Deer Flat Ranch at Big Sur. When at the ranch, Pauling was cared for in an unofficial capacity by his scientific colleague, Matthias Rath. Pauling was first visited by Rath, a physician, in 1989, having met him years earlier in Germany while on a peace tour. Rath was also interested in vitamin C, and Pauling took him on as a researcher at the Institute. There, the duo collaborated on investigations related to the influence of lipoproteins and vitamin C on cardiovascular disease.

Not long after Pauling's cancer diagnosis, a professor at UCLA, Dr. James Enstrom, published epidemiological studies showing that 500 mg doses of vitamin C could extend life by protecting against heart disease and also various cancers. This caused a resurgence of interest in orthomolecular medicine, and it seemed that Pauling and Rath's vision for the future of the Institute was looking brighter.

But as it happened, this bit of good news proved to be too little and too late. LPISM had already begun to disintegrate financially, its staff cut by a third. The Institute's vice president, Rick Hicks, resigned his position, and Rath, as Pauling's protégé, was appointed in his place. Following this, the outgoing president of LPISM, Emile Zuckerlandl, was succeeded by Pauling's eldest son, Linus Pauling Jr. Finally Pauling, his health in decline, announced his retirement as chairman of the board and was named research director, with Steve Lawson appointed as executive officer to assist in the day-to-day management of what remained of the Institute. (See Chapter 20 for further details.)

One day prior to his retirement as board chairman, Pauling signed a document in which he requested that Rath carry on his "life's work." Linus Jr. and Lawson, however, had become concerned about Rath's role at the Institute, focusing in particular on a patent transfer agreement that Rath had neglected to sign. Adhering to the patent document was a requirement for every employee at the Institute, including Linus Pauling himself. When pressed on the matter, Rath opted to resign his position, and was succeeded as vice president by Stephen Maddox, a fundraiser at LPISM.

After this transition, Pauling met with Linus Jr. to discuss the Institute's dire straits. Pauling's youngest son, Crellin, had also become more active with the Institute as his father's illness progressed, in part because he had been assigned the role of executor of Pauling's will. Together, Crellin, Linus Jr., and Lawson pieced together a path forward for LPISM. The group ultimately decided that associating the Institute with a university, and focusing its research on orthomolecular medicine as a lasting legacy to Pauling's work, would be the most viable avenue for keeping the organization alive. The decision to associate with Oregon State University, Pauling's undergraduate alma mater, had not been made by the time that Pauling passed away.

*

In February 1992, Pauling publicly announced that he had cancer. His critics responded with sentiments that were, at times, distinctly unsympathetic. In their view, Pauling's diagnosis undermined his decades-long advocacy of vitamin C as a preventative cancer treatment. Pauling retorted that most elderly men develop hyperplasia or cancer in their prostates, often by age 70, and he believed that it was quite likely, although not provable, that his high intake of vitamin C had delayed the inevitable by decades.

As Pauling continued to struggle with the limitations that his illness placed upon him, his new caretaker, ranch hand Steve Rawlings, became an important part of his life. Rawlings did a lot of the day-to-day work of providing for Pauling while he was ill, a time period during which Pauling increasingly sought out the solace and solitude of his isolated home on the Pacific Coast. In a 2011 interview, Steve Lawson would reflect on the importance of Pauling's Big Sur residence during the last few years of his life:

"When he was in Palo Alto, Pauling's time was sought by many people for many different reasons: old friends, colleagues, the public, the media. When he retreated to Deer Flat Ranch, he removed himself from that. I think he really loved that time alone down there. I know that he liked to watch some programs on T.V., some serialized programs. He read quite a lot. He loved to read mystery books. He was a rare individual in that there was really no division between what he did recreationally and what he did professionally. He was a scientist through and through, and derived pleasure from working on scientific problems. Often times if you go into someone's bathroom, you'll find a *Prevention* magazine, a *Reader's Digest*, or *Entertainment Weekly*, or *Time*, or the newspaper. Pauling's bathroom was stacked with scientific journals. He wasn't physically vigorous [by the early 1990s], but he certainly didn't seem fatigued."

With his time becoming increasingly precious, Pauling's coworkers, friends, and family all felt that he should do what he most wanted to do with his days, and this had always been to focus on science. Freed from the responsibility of running LPISM's day-to-day operations, Pauling continued to work at Deer Flat Ranch in spite of his worsening health problems.

Of particular interest was the fact that, stricken with cancer himself, Pauling's scientific fascination with the disease only intensified. Indeed, rather than remove himself from ongoing cancer research as his disease advanced, he instead committed even more fully to this cause in his final years. In particular, Pauling became increasingly interested in non-toxic methods of cancer therapy; methods, in other words, that were far less stressful on the body than are radiation or standard chemotherapy regimens.

In a paper co-authored with David Knight and Abram Hoffer, Pauling worked to determine survival rates among over 2,000 cancer patients receiving high doses of vitamin C and other nutrients. He even flew to Tulsa, Oklahoma in October 1992 for a conference on alternative treatments of cancer. He likewise continued his efforts to convince American physicians of the value of vitamin C and lysine in preventing and treating heart disease, a notion that was gradually beginning to gain small slivers of recognition in the medical community.

More privately, Pauling was waging a personal war on his disease, exploring avenues of immunotherapeutic treatment that were unorthodox in the medicine of the time but which have, in recent years, begun to show great promise. In a letter to medical writer and cancer consultant Ralph Moss, Pauling detailed a therapy involving autologous anticancer antigen preparation, or AAAP, of which he was somewhat skeptical but nonetheless interested in pursuing. Working with friends and colleagues at Stanford Medical School to raise monoclonal antibodies against his prostate cancer cells, Pauling ultimately conducted what amounted to exploratory and self-experimental science to try and discern the potential value of AAAP.

Pauling's first exposure to the idea of AAAP came from the work of Duncan McCollester, a medical doctor based in Irvington, New York, who advocated for a form of "Active Specific Immunotherapy." This treatment involved the use of a manganese phosphate gel that was mixed with isolated portions of tumor tissue in which tumor antigens had been converted to a form capable of stimulating a cancer-destroying immune response, once readministered in a patient's forearm or thigh. McCollester dedicated a book on the subject to Pauling, even as he was struggling to receive FDA approval for the treatment.

David Stipp, a reporter for the *Wall Street Journal*, wrote in August 1992 that a similar medical treatment had been developed by Cellcor, Inc. of Newton, Massachussetts. Cellcor offered customers a treatment for kidney cancer in which a patient's own white blood cells were extracted, treated in such a way as to make them attack tumor cells, and then reintroduced into the patient's cells. Known as autolymphocyte therapy, or ALT, the treatment had been available commercially in Atlanta, Boston, and Orange County, California since around 1990. However, at the time, medical officials disputed the efficacy of Cellcor's anticancer therapy, arguing that not enough data had been collected to substantiate the company's claims.

By July 1992, Pauling had decided to move forward with AAAP treatment, the ultimate goal being a vaccine that would combat his own illness while also providing useful data for the science of the future. Subsequently, a 1-gram section of cancerous tumor tissue that had been surgically removed from Pauling's body was shipped to McCollester's lab in New York. Upon entering the operation, Pauling's surgeon had advised him that his entire tumor should be removed, rather than a small section. Pauling refused this request, arguing that a full resection would prevent him and others from observing the effectiveness of the AAAP treatment. In other words, rather than focusing on the fact that his own life was on the line, Pauling was still operating, first and foremost, in the mode of the scientist: he was running an experiment in which he himself was a test subject, and the stakes could not have been higher. In Pauling's mind there were plenty of reasons for optimism.

By October, scientists at Stanford University, led by Ronald Levy, had successfully boosted the immune systems of a small group of B-cell lymphoma patients using a vaccine that had been genetically engineered from the patients' own tumor tissues. In two of these nine patients, tumors vanished completely.

In generating the vaccines for each individual patient, the Stanford scientists created clones of the cancerous B-cells from each subject, and then separated out specific proteins — known as receptors — from the outer coatings of the B-cells. Using genetic engineering techniques, the scientists then added other proteins that boosted the immune system and created a synthetic version of the engineered receptors. The result was a tailor-made vaccine created from the B-cell receptors that used each patient's immune system to attack cancer cells.

Pauling's confidence in his anticancer antigen treatment was further elevated by other immunotherapy techniques then being developed by a team at the National Cancer Institute, as headed by Steven Rosenberg, French Anderson, and Michael Blaese. However, these studies were all specific to skin cancer, and were carried out on patients already in remission and receiving chemotherapy, or on patients with very small tumors. Pauling, by contrast, was already afflicted with advanced prostate cancer by the time that his condition was discovered, and he had not yet accepted any form of radiation therapy.

From October 1992, Pauling almost exclusively used the AAAP vaccine and vitamin C to treat his cancer. The vaccine traveled from McCollester's lab in New York to a willing physician in California who had agreed to administer the suggested injections — anywhere from 0.2 to 0.65 mL a few times monthly. Pauling continued to receive these injections, which routinely caused tenderness and swelling, until January 1994, about seven months before he died.

After a sigmoidoscopy in 1993 revealed that his rectal tumor was still growing, the reality set in that he was not likely to survive his cancer. It was at this point that Pauling began to seriously consider which of his possessions should be turned over to family and which should be transferred to his archival collection at Oregon State University. The same year, it was decided that it would be a good idea to arrange a special symposium, co-sponsored by Caltech and LPISM, on or near his

93rd birthday. Speakers would consist of former graduate students and postdocs. Pauling had once imagined that an event of this sort would be appropriate for his 100th year, a birthday that he had fully intended to celebrate.

Throughout 1993, Pauling strived to be as active as possible, giving interviews in person or over the telephone, and entertaining many visitors at Deer Flat Ranch. At the end of May, Pauling and a collection of friends, family, and coworkers also gathered to celebrate the Institute's 20th anniversary. But as the months passed and his illness worsened, Pauling attended to his scientific writing and correspondence at a decreasing rate. On two occasions, he returned to Palo Alto to attend scientific meetings, giving a short talk at one. The last scientific paper that he authored himself was written in November-December of 1993. Much of his time was taken up with scheduled visits to his doctors in San Luis Obispo and Cambria, or simply resting at his sanctuary on the Pacific Ocean.

In January 1994, Pauling's physicians decided that steps needed to be taken to shrink his tumor, and Pauling relented to a course of chemotherapy, during which he attributed his lack of negative side effects to his routine ingest of vitamin C megadoses. When Pauling learned that the cancer had spread to his liver however, his hope to live to be 100 years old were lost. He stopped taking vitamin C completely, and gave up writing in his research notebook — a brief note about his work on nuclear structure appears in January and the pages after it are blank.

During the last months of his life, Pauling met with friends and family, while also attending to some less pleasant business. Steve Lawson and Linus Pauling Jr. journeyed to Deer Flat Ranch during this time to mediate ongoing litigation between the Institute and Matthias Rath, who had initiated a lawsuit against his former employer. Even at the deposition, which was given from his bed, Pauling welcomed Rath warmly.

Pauling's final public appearance came on June 19, 1994, at the conference that he had requested be organized a year earlier, and which his son Crellin had arranged. This event, which was ultimately hosted by The Pacific Division of the American Association for the

Advancement of Science, was titled "A Tribute for Linus Pauling" and was held at San Francisco State University. Pauling's ranch hand Steve Rawlings attended as Pauling's nurse, bringing him into the assembly in a wheelchair. Upon entering however, Pauling stood and insisted on walking into the room, receiving applause from the gathering as he made his way to his seat. An array of speakers including Harden McConnell, Alexander Rich, Frank Catchpool, Richard Kunin, and Crellin Pauling delivered moving talks that detailed Pauling's major contributions to science, human health, and world peace.

Pauling's daughter Linda was at Deer Flat Ranch with her husband and children on August 18, 1994, when Pauling suffered a stroke that left him comatose. Crellin and Linus Jr. arrived soon after and were both at the ranch with him on the evening that he died. His passing came at the end of a beautiful summer day, as the sun was just beginning to set over the Pacific. At the end of his life, Pauling wore on his wrist an opal bracelet that he had once given to Ava Helen as a gift.

In Palo Alto, Steve Lawson had just sat down for dinner when he received a call from Linus Jr., informing him of the sad news. Immediately, Lawson got in his car and went back to the Institute, faxing pre-written obituaries to the media. Copies went to CBS, the *New York Times*, NBC, CNN, the *San Francisco Chronicle*, the *San Jose Mercury-News*, and half a dozen more outlets. But by the time that Lawson had faxed the third news organization, the phone started ringing. He later recalled,

> "In those days, we had an old-fashioned phone system where you could see a number of little pegs that would light up for an incoming line, and I think there were as many as six incoming lines. Before long every light was lit and blinking: it was the *New York Times*, it was CBS, it was everybody under the sun that wanted statements."

Pauling's passing was reported the next day through packages of stories in the *New York Times* and the *Los Angeles Times* that were immediately picked up by news services for syndication around the globe. The *Pasadena Star-News* ran its own article a few days later, as did the *Medical Tribune* and the scientific journal *Nature*. Personal letters flooded in to the Institute and the Pauling children from every corner of the globe: France, the United Kingdom, Russia, Japan, Italy, Australia, South America, the Philippines, and all across the United States. Universities and organizations worldwide, including Caltech and the American Association for the Advancement of Science, sent heartfelt letters conveying their sadness at the loss of a great man.

In the months and years that followed, Pauling's life was honored in ways both large and small. The Alpha Chi Sigma chemistry fraternity, which is based in Indianapolis, dedicated the Library Room of its house to Pauling. A fossil leaf from an extinct species of citrus tree was also named after him: *Linusia paulinga*.

Later on in 1994, not long after Pauling's death, Lawson at LPISM received unmarked packages in the mail, containing nearly exact replicas of Pauling's Nobel Prizes. A week or two after they had arrived, Pauling's son Peter, who lived in Wales, called and cryptically asked if he had received anything "unusual" in the mail. As it turned out, Peter had gone to the Nobel Academies and had duplicate medals struck in an alloy for family members and the Institute to hold as keepsakes.

Later still, with the help of Linda Pauling and officials at Oregon State University, Lawson and others planned a Linus Pauling exhibition, which was sponsored by a Japan-based peace organization, Soka Gakkai International. Aiming to educate the public about Pauling's work and to introduce school children to Pauling as a role model, the exhibit focused on all facets of Pauling's career as a humanitarian, activist, scientist, and medical researcher. Over the course of several years, millions of people visited the exhibit in Europe, Japan, and many locations in the United States, including Washington D.C., San Francisco, and Boston. The exhibit was created by a team of designers who, when it had finished touring, donated all of its elements and infrastructure to Oregon State University.

*

On August 29, 1994, a memorial service planned by Pauling's children and his long-time assistant Dorothy Munro was held at Memorial Church on the campus of Stanford University. Many people spoke, including Steve Lawson, Oregon State University president John Byrne, and scientific colleagues Frank Catchpool and Verner Schomaker.

. Remembrances were likewise offered by close friends and family. Youngest son Crellin spoke movingly, while also offering comments written by his brother Peter, who was unable to make the trip from the U.K. Daughter Linda and eldest son Linus Jr. also gave their heartfelt goodbyes to their father. Steve Rawlings spoke of the bond that he and Pauling had formed during their time together at Deer Flat Ranch. Four of Pauling's grandchildren — Cheryl and David Pauling, and Barky and Sasha Kamb — recalled fond memories of their Grandpa.

The memorial program featured a quote to remember Pauling by, one taken from his 1958 book, *No More War*. It read:

> "Science is the search for truth — it is not a game in which one tries to beat his opponent, to do harm to others. We need to have the spirit of science in international affairs, to make the conduct of international affairs the effort to find the right solution, the just solution of international problems, not the effort by each nation to get the better of other nations, to do harm to them when it is possible. I believe in morality, in justice, in humanitarianism."

Before he died, Pauling made clear his wish to be cremated and to have his ashes, along with those of his wife, interred in Lake Oswego, Oregon at the Pioneer Cemetery where his parents were buried. In 1994, a cenotaph — which is a marker honoring a person whose remains are elsewhere — was placed in the family plot by Pauling's sister, Pauline. Pauling's ashes remained with Ava Helen's among family in California until 2005, when they were finally moved to Oregon and placed alongside those of Pauling's parents, Herman and Belle.

In 2013 an Oregon resident named Jean Crellin Ashby took her mother to see Linus Pauling's grave at Pioneer Cemetery. Ashby is the granddaughter of Edward Webster Crellin, once a major financial supporter of Pauling's at Caltech, and the man after whom Pauling named his youngest son. Standing over Pauling's marker, Ashby thought about how her grandparents were buried in Pasadena. Since she was unable to easily visit their graves, given the considerable distance, Ashby decided that honoring Pauling's family in Lake Oswego would also serve to honor her own. Subsequently, Ashby contacted cemetery administrators and filed the appropriate paperwork to become the official caretaker for the Pauling plot.

In reflecting on the end of Pauling's life, it seems appropriate to return to the diary entry that he wrote when he began recording his experiences at the age of 16. In it, the young Pauling wrote that his history was not intended to be merely a life's story. Rather, it was to be a reflection on good times had in his passage through this "vale of tears":

> "Often, I hope, I shall glance over what I have written before, and ponder and meditate on the mistakes that I have made — on the good luck that I have had — on the carefree joy of my younger days; and pondering, I shall resolve to remedy my mistakes, to bring back my good luck, and to regain my happiness."

Sourcing

Nearly all of the work presented in this book relied upon two second-ary sources to provide basic context: *Force of Nature: The Life of Linus Pauling* by Thomas Hager, and *The Pauling Chronology* by Robert Para-dowski. From there, original research was conducted primarily in the Ava Helen and Linus Pauling Papers (MSS Pauling) at the Oregon State University Libraries Special Collections and Archives Research Center (SCARC), with primary attention paid to the components of the collec-tion identified below.

Writing *The Nature of the Chemical Bond*

- Published in six parts (July 22 to August 27, 2014) to commemorate the 75th anniversary of the publication of *The Nature of the Chemical Bond*.
- Image: MSS Pauling: Series 5, Manuscripts and Typescripts of Books; Subseries 4, *The Nature of the Chemical Bond*. Box-Folder 4.001.1, Advertisement for *The Nature of the Chemical Bond*, 1939.
- Major sourcing:
  - MSS Pauling: Series 5, Manuscripts and Typescripts of Books; Subseries 4, *The Nature of the Chemical Bond*. Boxes 4.001 and 4.002 focus specifically on the first edition of Pauling's book.
  - MSS Pauling: Series 16, Personal Safe; Subseries 1, Linus Paul-ing's Safe — Drawer 1. This component of the collection contains nearly all of Linus Pauling's correspondence with his wife, Ava Helen. At some point after her death in 1981, Linus burned the majority of Ava Helen's letters to him, particularly those from

their early years together. Correspondence from the mid- to late-1930s remains extant in his personal safe.

Pauling's Battle with Glomerulonephritis

- Published in five parts from March 9 to March 23, 2010.
- Image: MSS Pauling: Series 9, Photographs. Item 1941i.8. Peter, Linus, Linus Jr., Linda, Crellin, and Ava Helen Pauling sitting on the front porch of their Pasadena home, 1941. Annotation: "1941, Daddy very ill". Photographer unknown.
- Major sourcing:
  - MSS Pauling, Series 4, Manuscripts and Typescripts of Speeches; Box-Folder 1941s.1 "The Structural Chemistry of the Future," March 7, 1941.
  - MSS Pauling: Series 1, Correspondence; Box 2, Addis, Thomas, 1940–1949.
  - MSS Pauling: Series 1, Correspondence; Box 3, Addis, Thomas, materials re: Pauling illness, 1941–1942.
  - MSS Pauling, Series 2, Publications; Box-Folder 1994p.6, "Thomas Addis (July 27, 1881–June 4, 1949)," by Kevin Lemley and Linus Pauling. *Biographical Memoirs (National Academy of Sciences)* 63 (1994): 2–46.

W.H. Freeman & Co.

- Published in eight parts from July 25 to September 12, 2018.
- Image: MSS Pauling: Series 5, Manuscripts and Typescripts of Books; Subseries 5, *General Chemistry*. Box-Folder 5.009.1, Advertisement for *General Chemistry*.
- Major sourcing:
  - MSS Pauling: Series 1, Correspondence; Box 439, W.H. Freeman & Company, 1941–1959.
  - MSS Pauling: Series 1, Correspondence; Box 440, W.H. Freeman & Company, 1960–1992, No Date.

o MSS Pauling: Series 1, Correspondence; Box-Folder 152.9, Hayward, Roger, 1951–1972. SCARC is also home to the Roger Hayward Papers (MSS Hayward).

Christmas

• Published on December 17, 2014, this recollection is excerpted from an oral history interview conducted with Linus Pauling Jr. by Chris Petersen on June 4, 2012, as cataloged into SCARC's History of Science Oral History Collection (OH 017).

• Image: MSS Pauling: Series 9, Photographs. Item 1925i.85, Linus Pauling Jr.'s first Christmas, 1925. Photographer unknown.

The Soviet Resonance Controversy

• Published in seven parts from May 27 to July 15, 2020.

• Image: MSS Pauling: Series 13, Biographical; Sub-Series 6, Scrapbook Pages Mounted by Linus Pauling. Box-Item 6.008.157 "Слава И Горлость Русской Науки" ["Glory and Pride of Russian Science"], *Pravda*, November 22, 1961.

• Major sourcing:

o *The Structural Chemistry of Linus Pauling*, by Robert J. Paradowski, 1972 (doctoral dissertation).

o MSS Pauling: Series 11, Science; Subseries 2, Quantum Mechanics; Box 2.003, Materials re: Resonance Theory Controversy, 1949–1983.

o MSS Pauling, Series 4, Manuscripts and Typescripts of Speeches; Box-Folder 1946s.3, "Theory of Resonance," February 8, 1946; Box-Folder 1961s3.17, "Resonance in Chemistry," December 3, 1961.

o MSS Pauling, Series 2, Publications; Box-Folder 1977p.8, "The Theory of Resonance in Chemistry," *Proc. Roy. Soc. Lond.* A356 (September 1977): 433–441.

    o The author of these posts, Miriam Lipton, is fluent in Russian and was able to engage with Russian-language materials that had previously been inaccessible to us as researchers.

One Year as President of the American Chemical Society

- Published in four parts (October 9 to October 30, 2019) to commemorate the 70th anniversary of Pauling's presidential year.
- Image: MSS Pauling: Series 13, Biographical; Sub-Series 6, Scrapbook Pages Mounted by Linus Pauling. Box-Item 6.005.020, Assembled images of Pauling ACS Presidency announcement.
- Major sourcing:
  - MSS Pauling: Series 11, Science; Subseries 14, Scientific, Research and Grant-Funding Organizations. Boxes 14.004 to 14.010 all document Pauling's connection with the ACS during his presidential year.
  - MSS Pauling: Series 2, Publications; Box-Folder 1949p.3, "Our Job Ahead," *Chem. Eng. News* 27 (January 1949): 9.
  - MSS Pauling: Series 4, Manuscripts and Typescripts of Speeches; Box-Folder 1949s.14, "Chemistry and the World of Today," September 19, 1949.
  - A definitive account of Ralph Spitzer's firing from Oregon State College is available in "The Academy and Cold War Politics: Oregon State College and the Ralph Spitzer Story," by William G. Robbins. *Pacific Northwest Quarterly* 104 (Fall 2013): 159–175.

The Theory of Anesthesia

- Published in five parts from June 2 to June 16, 2009.
- Image: MSS Pauling: Series 9, Photographs. Item 1963i.26, Linus Pauling shaking hands with Frank Catchpool at a Caltech event, 1963. Photographer unknown.
- Major sourcing:
  - MSS Pauling: Series 11, Science; Subseries 12, Anesthesia. Box 12.001.

- o MSS Pauling: Series 2, Publications; Box-Folder 1961p.8, "A Molecular Theory of General Anesthesia," *Science* 134 (July 1961): 15–21.

Lloyd Jeffress

- Published July 2, 2009.
- Image: MSS Pauling: Series 1, Correspondence; Box-Folder 189.1, Jeffress, Lloyd A., 1923–1991. Lloyd Jeffress letter envelope with Pauling annotations.
- Major sourcing:
  - o MSS Pauling: Series 1, Correspondence; Box-Folder 189.1, Jeffress, Lloyd A., 1923–1991.
  - o MSS Pauling: Series 3, Manuscripts and Typescripts of Articles; Box-Folder 1986a.7, "Life with Lloyd Jeffress," June 5, 1986.

Deer Flat Ranch

- Published in three parts from September 21 to October 5, 2016.
- Image: MSS Pauling: Series 9, Photographs. Item 1963i.37 Linus Pauling standing outside, harvesting abalone, July 1963. Photograph by Linus Pauling Jr.
- Major sourcing:
  - o MSS Pauling: Series 13, Biographical; Sub-Series 4, Business and Financial. Boxes 4.046 to 4.053 document the design, construction and management of Deer Flat Ranch.
  - o MSS Pauling: Series 13, Biographical; Sub-Series 5, Personal and Family; Box 5.057, Notebooks re: Pauling Family History and Financial Records.
  - o MSS Pauling: Series 6, Research Notebooks; RNB 7, 23, 28 and 35.
  - o OH 017: Oral history interviews of Linus Pauling Jr. and Stephanie Pauling conducted by Chris Petersen in June 2012.

Stuck on a Cliff

- Published June 25, 2009.
- Image: MSS Pauling: Series 13, Biographical; Sub-Series 6, Scrapbook Pages Mounted by Linus Pauling. Box-Item 6.007.521, "Pauling Saved From Coast Cliff", *New York Times*, February 1, 1960.
- Major Sourcing:
  - o MSS Pauling: Series 16, Personal Safe; Sub-Series 2, Linus Pauling's Safe — Drawer 2; Item 25.2, Letter from LP to children re: his experience on a cliff, February 12, 1960.
  - o MSS Pauling: Series 13, Biographical; Sub-Series 6, Scrapbooks Mounted by Linus Pauling; leaves 7.503 to 7.562 document the "cliff incident" in detail.
  - o MSS Pauling: Series 7, Newspaper Clippings; Box 1960n.
  - o MSS Pauling: Series 1, Correspondence; Box-Folder 31.1, Brando, Marlon, 1960.

The SISS Hearings

- The SISS hearings were covered in two batches of posts that were published in commemoration of the 50th anniversary of Pauling's appearances before the Senate subcommittee. The June hearing was covered in five posts released June 15 to June 29, 2010, and the October hearing was covered in a three-part series published October 13 to October 27, 2010.
- Image: MSS Pauling: Series 13, Biographical; Sub-Series 2, Political Issues. Box-Folder 2.021.41, "Pauling Again Refuses Names to Senate Quiz", *Los Angeles Times*, October 12, 1960.
- Major sourcing:
  - o MSS Pauling: Series 13, Biographical; Sub-Series 2, Political Issues. Boxes 2.012 to 2.024 document Pauling's interactions with the SISS.
  - o MSS Pauling, Series 4, Manuscripts and Typescripts of Speeches; Box-Folder 1960s2.7, "Our Need for Peace and Disarmament,"

July 9, 1960; Box-Folder 1960s3.9, "Civil Liberties in the Nuclear Age," November 19, 1960.

o Seven different typescripts at least partially related to the SISS are contained in MSS Pauling: Series 3, Manuscripts and Typescripts of Articles; Box 1960a2.

Unitarianism

- Published July 29, 2010.
- Image: MSS Pauling: Series 9, Photographs. Item 1969i.24, Stephen Fritchman and Linus Pauling standing on a walkway by the beach, leaning against a low brick wall. La Jolla, California, 1969. Photographer unknown.
- Major sourcing:
  - o MSS Pauling: Series 1, Correspondence; Box-Folder 121.11, Fritchman, Stephen H., 1950–1981.
  - o MSS Pauling: Series 15, Ava Helen Pauling; Sub-Series 2, Manuscripts and Typescripts of Articles and Speeches; Box-Folder 2.003.1, "Why I am a Unitarian," September 18, 1977.
  - o MSS Pauling, Series 4, Manuscripts and Typescripts of Speeches; Box-Folder 1969s.19, "Stephen Fritchman and the First Unitarian Church of Los Angeles," December 28, 1969.

An Honorary Diploma from Washington High School

- Published July 20, 2012 and updated for inclusion in this book.
- Image: MSS Pauling: Series 9, Photographs. Item 1962i.42, Linus Pauling smiling and pointing at his Washington High School honorary diploma, 1962. Photographer unknown.
- Major sourcing:
  - o MSS Pauling: Series 1, Correspondence; Box-Folder 441.1, Washington High School, 1947–1981.
  - o MSS Pauling: Series 8, Honors, Awards, Citations, Diplomas and Other Recognitions; Box-Folder 1962h.2, Washington High

School, Honorary high school diploma, [Includes correspondence]. June 14, 1962.

o "Washington High School's Haunted Halls," *Portland Mercury*, August 20, 2009; "Washington High School Development to Include Music Venue," *Portland Mercury*, July 11, 2014.

Receiving the Nobel Peace Prize

- Published in commemoration of the 50th anniversary of Pauling's receipt of the Nobel Peace Prize, this six-part series was released over the course of Fall 2013, with posts appearing on October 10, November 6 and 13, and December 4, 10 and 17, 2013.
- Image: MSS Pauling: Series 9, Photographs. Item 1963i.20, Linus Pauling sitting and eating a meal, Ava Helen Pauling standing above him holding a drink in hand. A banner on the wall reads "Paulings Put Pace in the Peace Race," October 1963. Photographer unknown.
- Major sourcing:
  o The authors of these posts made effective use of the data presented in SCARC's Linus Pauling Day-by-Day web resource to tease out a great many sources that detailed the prelude to and granting of the Nobel Peace Prize in Fall 1963. See: http://scarc.library.oregonstate.edu/coll/pauling/calendar/1963/index.html.
  o Pauling's documentation of the story behind his being awarded the Nobel Peace Prize is recorded in MSS Pauling: Series 6, Research Notebooks, RNB 23.
  o For more on the Pauling-Teller Debate, see MSS Pauling: Series 2, Publications; 1958p2.1, "Fallout and disarmament: A debate between Linus Pauling and Edward Teller." [Debate held on February 20, 1958 in San Francisco and broadcast on KQED] *Daedalus: Proc. Am. Acad. Arts and Sciences* 87 (1958): 147–163.
  o More about the *Life* magazine editorial "A Weird Insult from Norway" is available in MSS Pauling: Series 13, Biographical; Sub-Series 3, Legal, Box-Folder 3.059.3.

- o Extensive documentation of the response to Pauling's Nobel Peace Prize as well as his Nobel trip are held in MSS Pauling: Series 8, Honors, Awards, Citations, Diplomas and Other Recognitions, boxes 1963h2 to 1963h6.
- o The five lectures that Pauling gave while in Norway and Sweden were as follows. MSS Pauling: Series 4, Manuscripts and Typescripts of Speeches, Box-Folder 1963s.21, "Response by Linus Pauling," Nobel Peace Prize Ceremony, Oslo, Norway, December 10, 1963; Box-Folder 1963s.22, "Science and Peace," Nobel Lecture, Oslo, Norway, December 11, 1963; Box-Folder 1963s.23, "Humanism and Peace," Nurmann Lecture, Trondheim, Norway, December 16, 1963; Box-Folder 1963s.24, No Title, [re: nations and morality; armament for peace], Stockholm, Sweden, December 19, 1963; Box-Folder 1963s.25, "Next Steps Toward Peace," Lund University, Lund, Sweden, December 21, 1963.

William P. Murphy

- Published April 2, 2009.
- Image: Photo of William Murphy extracted from "Success Story: How an Oregon Man Won the Nobel Prize," *The Sunday Oregonian*, July 21, 1935.
- Major sourcing:
  - o MSS Pauling: Series 1, Correspondence; Box-Folder 252.4, Murphy, William P., 1983.
  - o "Success Story: How an Oregon Man Won the Nobel Prize," *The Sunday Oregonian*, July 21, 1935.

The *National Review* Lawsuit

- A two-part series published on January 30 and February 7, 2013.
- Image: MSS Pauling: Series 13, Biographical; Sub-Series 6, Scrapbook Pages Mounted by Linus Pauling. Box-Item 6.008.407,

"National Review Sued for Million by Pauling over Traitor Label", *New York Times*, May 21, 1963.

- Major sourcing:
  - o MSS Pauling: Series 13, Biographical; Sub-Series 3, Legal. Boxes 3.008 to 3.017 focus on Pauling v. *National Review*, Buckley, et al.

Peter Pauling

- Published in nine parts from May 11 to July 6, 2016.
- Image: MSS Pauling: Series 9, Photographs. Item 1952i.5, Crellin, Linda, and Peter Pauling standing outside, 1952. Photographer unknown.
- Major sourcing:
  - o MSS Pauling: Series 13, Biographical; Sub-Series 5, Personal and Family. Boxes 5.041 to 5.046 contain Peter Pauling's correspondence with his parents and siblings, as well as his childhood schooling records. Box 5.047 contains correspondence authored or received by his wives and children. This sub-series also contains five boxes of lightly processed materials related to Peter Pauling's academic and professional work.
  - o "The Personal View of Linus Pauling and His Work," by Crellin Pauling. In *The Pauling Symposium: A Discourse on the Art of Biography. Proceedings of the Conference on the Life and Work of Linus Pauling (1901–1994)*, Ramesh Krishnamurthy, Clifford Mead, Mary Jo Nye, Sean Goodlett, and Marvin Kirk, eds. Corvallis, Oregon: Special Collections, Oregon State University Libraries, 1996.
  - o MSS Pauling: Series 11, Science; Sub-Series 9, The Nucleic Acids, The Structure of Atomic Nuclei; Box 9.001, Nucleic Acid Papers, 1951–1963. Note in particular item 1.44, the Francis Crick "hoax letter" forged by Peter Pauling.
  - o OH 017: Oral history interviews of Stephen Lawson and Linus Pauling Jr. conducted by Chris Petersen between August 2011 and June 2012.

Patents

- The original patents series was published in installments from September 19, 2012 to January 23, 2013; 13 posts in total. For this book, an additional post providing needed context on the Pauling Oxygen Meter was included. That post was originally published on November 23, 2010.
- Image: MSS Pauling: Series 16, Personal Safe; Subseries 1, Linus Pauling's Safe — Drawer 1. Box-Item 31.7, Diagram of the Pauling Oxygen Meter included with letter from Clarence F. Kiech to Linus Pauling, Reuben Wood, and James H. Sturdivant, June 8, 1942.
- Major sourcing:
  o MSS Pauling: Series 11, Science; Sub-Series 10, LP Patents, LP Notes to Self, Other Fields of Science. Boxes 10.001 to 10.002 focus on Pauling's patents and invention submissions.
  o MSS Pauling: Series 11, Science; Sub-Series 13, Scientific War Work. Boxes 13.001 to 13.003 focus on the Pauling Oxygen Meter.
  o Interviews with Linus Pauling and Joseph Koepfli by Thomas Hager as held at SCARC in the Thomas Hager Papers (MSS Hager): Series 2, *Force of Nature*: Interviews — Audio Cassettes and Transcriptions.
  o MSS Pauling: Series 2, Publications; Box-Folder 1989p.8, "Explanations of cold fusion," *Nature* 339 (May 1989): 105.
  o OH 017: Oral history interview of Stephen Lawson conducted by Chris Petersen, December 15, 2011.

Jack Kevorkian and a Twist on Capital Punishment

- Published June 3, 2011.
- Image and major sourcing: MSS Pauling: Series 1, Correspondence; Box-Folder 198.7, Kevorkian, Jack, 1973, 1976.

The Linus Pauling Institute of Science and Medicine

- Marking the Linus Pauling Institute's 40th anniversary, this series was published in nine parts from March 13 to May 8, 2013.
- Image 1: MSS Pauling: Series 9, Photographs. Item 1989i.62, Linus Pauling Institute of Science and Medicine group portrait, January 1989. Photographer unknown.
- Image 2: Portrait of Linus Pauling Jr. by Christy Turner, Oregon State University Libraries.
- Major sourcing:
  - MSS Pauling: Series 1, Correspondence. Boxes 223 to 227 contain records related to the Linus Pauling Institute of Science and Medicine; box 222 is devoted to the Linus Pauling Heart Foundation. SCARC is also home to the voluminous Linus Pauling Institute of Science and Medicine Records (MSS LPISM), though this collection had not been processed when these blog posts were being researched.
  - OH 017: Oral history interviews of Stephen Lawson and Linus Pauling Jr. conducted by Chris Petersen between August 2011 and June 2012.
  - Linus Pauling Jr.'s October 2011 keynote address has been cataloged into MSS Pauling: Series 13, Biographical; Sub-Series 5, Personal and Family; Box 5.039, Family Correspondence, Linus Carl Pauling Jr., 1975–1996, No Date.

A Funny Story from Andy Warhol

- Published April 6, 2016. Warhol's story originally appeared in *The Andy Warhol Diaries*, authored by Warhol and edited by Pat Hackett. New York, NY: Warner Books, 1989.
- Image: MSS Pauling: Series 9, Photographs. Item 1985i.48, Linus Pauling, Jill Sackler and Andy Warhol, 1985. Photographer unknown.

<u>Quasicrystals</u>

- Published in four parts from May 2 to May 23, 2012.
- Image 1: MSS Pauling: Series 9, Photographs. Item 1986i.48, Linus Pauling speaking in 1986. Photographer unknown.
- Major sourcing:
  - o MSS Pauling: Series 11, Science; Sub-Series 4, The Structure of Quasicrystals, Superconductivity. Boxes 4.001 to 4.005 focus on Pauling's quasicrystals work.
  - o MSS Pauling: Series 2, Publications; Box-Folder 1987p.1, "So-called icosahedral and decagonal quasicrystals are twins of an 820-atom cubic crystal," *Phys. Rev. Lett.* 58 (January 1987): 365–368; Box-Folder 1987p.13, "Evidence from X-ray and neutron powder diffraction patterns that the so-called icosahedral and decagonal quasicrystals of MnAl6 and other alloys are twinned cubic crystals," *Proc. Natl. Acad. Sci.* 84 (1987): 3951–3953; "'The nonsense about quasicrystals,'" *Science News* 129.1 (January 4, 1986): 3. [No item-level cataloging in MSS Pauling.] Pauling published at least 15 papers on quasicrystals from 1985 to 1991.
  - o "My Memories and Impressions of Linus Pauling," by David Shoemaker. In Krishnamurthy, et al., 1996.
  - o This very technical set of posts benefitted from the assistance of OSU Chemistry professor Arthur Sleight, who provided feedback as the content was being drafted. The original posts were also enhanced by animated gifs created by OSU student Geoff Bloom.

<u>Pauling's Last Years</u>

- Published in four parts from July 27 to August 17, 2016.
- Image: MSS Pauling: Series 9, Photographs. Item 1994i.8, Linus Pauling speaking at the International Symposium on Biological NMR, Stanford University, March 1994. Photographer unknown.

- Major sourcing:
  - o MSS Pauling: Series 13, Biographical; Sub-Series 1, Academia; Box-Folder 1.001, "Diary (so-called)", by Linus Pauling, 1917–1918.
  - o MSS Pauling: Series 13, Biographical; Sub-Series 5, Personal and Family. Boxes 5.060 to 5.062 document Pauling's final illness as well as his passing, funeral and marker at Oswego Pioneer Cemetery.
  - o MSS Pauling: Series 6, Research Notebooks. Pauling chronicled his cancer treatment in several autobiographical entries logged in RNB 44.
  - o MSS Pauling: Series 7, Newspaper Clippings. Box 1992n contains news coverage of Pauling's cancer diagnosis; boxes 1994n to 1995n5 contain Pauling's published obituaries.
  - o OH 017: Oral history interview of Stephen Lawson conducted by Chris Petersen on September 19, 2011.

Other Images

- Cover: MSS Pauling: Series 9, Photographs. Item 194-i.012, Studio portrait of Linus Pauling, ca. 1940s. Photographer unknown.
- Preface: Photo by Philip Vue, Oregon State University Libraries.
- Biographical Sketch: MSS Pauling: Series 9, Photographs. Item 1918i.29. Linus Pauling, age 17. Tillamook, Oregon. 1918. Photographer unknown.

Index

Printed in the United States
by Baker & Taylor Publisher Services